262  **World Bank Discussion Papers**

Public Policy for the Promotion of Family Farms in Italy

The Experience of the Fund for the Formation of Peasant Property

Eric B. Shearer
Giuseppe Barbero

The World Bank
Washington, D.C.

Copyright © 1994
The International Bank for Reconstruction
and Development/THE WORLD BANK
1818 H Street, N.W.
Washington, D.C. 20433, U.S.A.

All rights reserved
Manufactured in the United States of America
First printing September 1994

Discussion Papers present results of country analysis or research that are circulated to encourage discussion and comment within the development community. To present these results with the least possible delay, the typescript of this paper has not been prepared in accordance with the procedures appropriate to formal printed texts, and the World Bank accepts no responsibility for errors. Some sources cited in this paper may be informal documents that are not readily available.

The findings, interpretations, and conclusions expressed in this paper are entirely those of the author(s) and should not be attributed in any manner to the World Bank, to its affiliated organizations, or to members of its Board of Executive Directors or the countries they represent. The World Bank does not guarantee the accuracy of the data included in this publication and accepts no responsibility whatsoever for any consequence of their use. The boundaries, colors, denominations, and other information shown on any map in this volume do not imply on the part of the World Bank Group any judgment on the legal status of any territory or the endorsement or acceptance of such boundaries.

The material in this publication is copyrighted. Requests for permission to reproduce portions of it should be sent to the Office of the Publisher at the address shown in the copyright notice above. The World Bank encourages dissemination of its work and will normally give permission promptly and, when the reproduction is for noncommercial purposes, without asking a fee. Permission to copy portions for classroom use is granted through the Copyright Clearance Center, Inc., Suite 910, 222 Rosewood Drive, Danvers, Massachusetts 01923, U.S.A.

The complete backlist of publications from the World Bank is shown in the annual *Index of Publications,* which contains an alphabetical title list (with full ordering information) and indexes of subjects, authors, and countries and regions. The latest edition is available free of charge from the Distribution Unit, Office of the Publisher, The World Bank, 1818 H Street, N.W., Washington, D.C. 20433, U.S.A., or from Publications, The World Bank, 66, avenue d'Iéna, 75116 Paris, France.

ISSN: 0259-210X

Eric B. Shearer and Giuseppe Barbero are consultants to the Agricultural Policies Division of the World Bank's Agriculture and Natural Resources Department.

Library of Congress Cataloging-in-Publication Data

Shearer, Eric B. (Eric Berthold), 1922–
    Public policy for the promotion of family farms in Italy : the
experience of the Fund for the Promotion of Peasant Property / Eric
B. Shearer and Giuseppe Barbero.
      p.   cm. — (World Bank discussion papers ; 262)
    Includes bibliographical references.
    ISBN 0-8213-3026-8
    1. Family farms—Government policy—Italy. 2. Cassa per la
Formazione della Proprietà Contadina.   I. Barbero, Giuseppe.
II. Title.   III. Series.
HD1476.I8S53   1994
    338.1'6—dc20   94-33267
                                           CIP

CONTENTS

* * *

*   *   *

Tables

FOREWORD

Land ownership redistribution and the establishment of an efficient, household-based, farming sector are important policy objectives on the agenda of many developing countries. This paper is part of a larger recent effort oriented towards a better understanding of how these policy objectives can be achieved in a sustainable way, relying to the extent possible on market forces. Experience with land reforms is not well known, and thus market-based information on the possibilities, prerequisites, and financial implications of such interventions is scarce. This study of the Italian experience since the end of World War II offers the opportunity to glean the lessons from the experience of a mature program, spanning several decades of operation. The study augments an emerging body of evidence on the potential and costs of market-assisted land reform. It will help development agencies and governments in formulating land reform strategies.

Michel Petit
Director
Agriculture and Natural Resources
Department

v

ABSTRACT

Italian public policy, as well as the European Community in general, especially since the end of World War II, has emphasized the socio-economic and political value of the family farm. Along with, and subsequent to, the regionalized land redistribution via expropriation in key depressed areas in the early 1950's, subsidized credit and fiscal relief have helped promote the formation or enlargement of hundreds of thousands of family-sized farms via the market place.

A special agency, the *Cassa per la Formazione della Proprieta Contadina*, was created to act as intermediary between potential sellers and buyers and as long-term, low-interest lender to about 20,000 individual buyers and 166 farming cooperatives.

The implicit capital grant element from 1948 to 1991 has fluctuated between 60 and 70 percent and the average grant element per client family has been equivalent to about US$220,000 at 1991 prices, largely owing to the negative real interest rate during a period of high inflation.

Actual and hypothetical case studies demonstrate how, despite the largely favorable "terms of trade" between land prices and farm wages since the 1960s, many borrowers would not have been able to maintain a standard of living above the poverty line if they were to have paid the full market price for both the land and the credit. This appears to validate the need for a grant element in support of the public strategy; the size of the grant element is a political decision based on economic and fiscal realities.

ACKNOWLEDGEMENTS

The authors wish to acknowledge, above all, the valuable contribution of our two research assistants, Pier Domenico Ceccaroni Cambi Voglia and Franco Quintieri. They mined the historical files of the *Cassa* and performed the systematic statistical analyses on the data they uncovered, which are reflected in Chapter 4. They also participated in the case studies and provided the first drafts on the basis of their field notes. In addition, they have been very useful in assisting with numerous other searches for, and analyses of, data.

Next, our thanks go to the President and Director General of the *Cassa*, who have kindly welcomed the study and opened the files, and in particular to the director of Technical Services, Mr. Giuseppe Buccolini, who gave generously of his time for frequent consultations and was our guide during a week's travel the length of the Italian peninsula visiting the case study farms that he had selected and made arrangements with. We also want to express our appreciation to the farm families as well as to the management staff of the cooperatives, all of whom received us so hospitably and forthrightly.

The authors also gratefully acknowledge the cooperation of Dr. V. Pontolillo (Bank of Italy), Dr. E. de Medio (National association of agricultural credit institutions - ANICA), Dr. M. Roccia (Italian banking association - ABI) and Dr.G. Coppola (former Ministry of agriculture and forestry).

The enthusiasm of Professor Lorenzo Venzi of the University of Tuscia and director of INEA's Observatory for Latium and Umbria, as well as his valuable comments on the final draft, are much appreciated.

Last but not least, the study owes its being to the active interest of Dr. Gershon Feder of the World Bank in rounding up funding for its realization. Moreover, he provided valuable guidance in methodology and focus by reviewing and discussing with us various working drafts.

Part One: Italy's Postwar Agrarian Problems and Remedial Action

Introduction: The postwar situation and policy decisions in favor of the small family farm

Less than two months before the first postwar Italian elections in early 1948, the provisional government issued two far-reaching decrees designed to facilitate the formation, by means of financial and fiscal incentives, of what was then called "small peasant property". The timing of the decrees obviously reflected the electoral climate of the moment. Indeed, the decrees were the immediate precursors of the two rather radical land reform laws pushed through the new Parliament by the Christian Democrats and their allies only two years later, as responses to the widespread land hunger of the moment among the poorest strata of the rural population.[1]

The nature of the policy change can be fully grasped only by recalling the problematic situation of many rural areas at the end of World War II in terms of population density, land tenure relations and disparities in incomes and social conditions.

There were 8.5 million *contadini*[2] of both sexes, representing over 42% of the total active population, on a useable agricultural area of about 18 million hectares, more than 60% of which was hilly or mountainous (1951 Census of agriculture). The few costly prewar public land reclamation and settlement programs had been totally inadequate to correct the man/land ratio (nor were they designed to do so). On the contrary, the still relatively high birth-rate in the rural areas (encouraged by the fascist regime) and the sharply limited opportunities for internal and external migration during the inter-war period had contributed to the building up of considerable population pressure on the land. In the North and most of Central Italy, the family farm with on-farm or nearby dwelling had become well-institutionalized -- often at considerable sacrifice by the family concerned -- on either owned or rented land or through crop-share arrangements. Not so in the South and the Islands and in the vast coastal plain of Latium and Tuscany known as the Maremma, where (despite some noteworthy exceptions) the most characteristic structure was still the extensively used *latifondo*, worked typically by day labor or under relatively insecure and short-term cash or share tenancy systems especially in Sicily, the large landowner would rent pieces of his estate to various intermediaries (*gabellotti*), who tended to represent an extreme form of exploitation of the sub-tenants or laborers.)

Rossi Doria (q.v.) showed that from 1923/1928 to 1951 the active farm population in the South had grown by one-fourth while the net product per capita had declined by 22%, as against an increase of similar magnitude in the North, where development had already reduced the farm population tangibly. Even as late as 1961, the density of farm-dependent population in the South

---

[1] In Italy, as well as in the rest of western Europe, a series of measures were adopted after the end of World War II, which can be considered a result at one and the same time of farm workers' becoming aware of their rights and of the will (or need) of the respective governments to be somehow responsive. They are essentially measures for the redistribution of incomes in favor of the agricultural sector and particularly in favor of the lowest income families ... (Barbero 1964, p. 380)

[2] The term -- which has fallen into disuse in modern Italy -- denoted a socio-economic category, distinct from the "cittadino", or city-dweller, in the first instance, characterized by traditional attachment to the soil, which was worked by hand and/or with animal power. It was applied equally to independent small farm operators, small tenants and share-croppers and hired hands and day laborers. Even the urban middle-class owner of a small farm parcel somewhere would speak of his "contadino" when referring to its caretaker or tenant.

hardly differed from what it had been in the 1920s, while per capita income was only half the North's average.

It was this situation of extreme contrasts and resulting social conflicts at the end of World War II (exacerbated by the destruction and hardships brought on by four years of war), that again brought the "agrarian question" to the fore among intellectuals and journalists and as battle cry of the politicized farm labor unions and the leftist parties (soon to be dominated by the Communists). As a response to these pressures, in the first years of the newly reconstituted democracy, several partial relief measures were enacted in the form of minimum farm employment coefficients on large estates, concessions of uncultivated or "insufficiently cultivated" estates to workers' cooperatives (sometimes in recognition of extra-legal occupations), freezing of rental and crop-share contracts and rent control. Meanwhile, heated debate continued over the implementation of Article 44 of the new Italian Constitution[3].

The large majority won in the first postwar political election in 1948 by the Christian Democrats and their allies over the "Popular Front" led to a gradual dismissal of the idea of a general landownership redistribution. This idea was killed *de facto* by Parliament's passage in 1950 of two laws providing for partial expropriation and redistribution (on easy terms) to *contadini* of large estates in the problem areas (Southern Italy, Sardinia, the Maremma and the Po Delta). Implementation of these laws was accompanied by enormous investments in land reclamation, irrigation and rural infrastructure (including housing), technical assistance, provision of working capital and development of processing cooperatives. The investments were channeled partly through the *Cassa per il Mezzogiorno* (Fund for the South), to which the Marshall Plan as well as the World Bank made important contributions. (A separate land reform law, passed by the Sicilian regional assembly, had far less redistributional and productive impact, although substantial funds were channelled to Sicily as well.)

The land reform laws, together with the two above-mentioned legislative decrees providing incentives for the spontaneous formation of peasant properties, and alongside the laws for tenancy regulation and rent freezing, represent the core measures of Italy's postwar agrarian policy. The policy can be defined as having had two main objectives:

(a) to eliminate or reduce, in the short run, the social and political conflicts that arose from inequitable resource distribution and power relations;

(b) to promote and assist the transformation of landless workers and poor *minifundia* owners, tenants and share croppers into a politically stable and moderate class of small but independent

---

[3]Article 44 of Italy's 1948 Constitution reads as follows (in translation):

Obligations and restrictions on private land ownership.

In order to attain the rational use of the soil and establish equitable social relations, the law imposes obligations and restrictions on private land ownership, fixes limits on its size according to regions and agrarian zones, promotes and imposes land reclamation, the transformation of the *latifondo* and the reconstitution of productive units; it lends assistance to small and medium sized property.

...

4

freeholders whose main concern was to be the management and improvement of their newly acquired land resources through the full utilization of family labor.

Indeed, only in recent years have the various types of tenancy contracts been *de facto* "liberalized", while land purchases by eligible farm families or cooperatives still enjoy substantial (though not unlimited) fiscal and financial subsidies. The special agency created in 1948 to implement a part of the latter, the *Cassa per la formazione della proprietà contadina* (Fund for the formation of peasant property) has, over time -- and especially during the last decade -- acquired an expanding role in promoting land mobility and further expansion of the family farm sector.

Structure of the paper.

Analysis of the nature, *modus operandi*, impact and costs of the *Cassa* is the main subject of this study. Its basic motivation stems from the idea that, to a greater or smaller extent, the Italian experience could be of interest to countries that are still faced with the problem of modernizing the agricultural sector while creating more equitable and stable social relations in the rural areas. Such an analysis must, of course, be related to the country's economic and social history. Furthermore, it cannot be separated from consideration of the general system of incentives that has been an important component of Italy's agricultural policy.

Thus, the first two chapters offer a summary of the overall measures and of the peasants' relation to the land market, within the context of Italy's postwar agrarian situation.

Chapter 3 provides an overview of the impact on the land market and tenure structure of the general fiscal and financial incentives enacted, respectively, in 1948 and 1965 and following the regional decentralization of certain government functions in the mid-1970s.

A brief analysis of farmer purchasing power in terms of land is attempted in Chapter 4.

Part Two of the paper deals with the *Cassa*. Its history, nature and operations are described in Chapter 5. Its impact and its resources and the public cost, in terms of the grant element represented by the subsidized interest rate, are analyzed in Chapter 6. A few hypothetical cases going back to earlier years are constructed on the basis of historical data.

Chapter 7 deals with a small number of case studies of recent clients of the Fund (i.e., not with the rural poor of the immediate postwar period); farm account data are used to show the incidence of the actual subsidy on family welfare, and to estimate the minimum level of the subsidy that would have sufficed to keep the family above the poverty line while paying off a 30-year loan on the market price of the property.

In the final chapter, the authors strive to draw a number of conclusions from the Italian experience for the consideration of researchers and policy formulators in developing countries -- particularly in Latin America -- and in the former centrally planned economies.

Chapter 2

The drive towards peasant property in Italy: Peasant property expansion in the inter-war years

Typically, the secure possession of a piece of farmland -- or of more land, as the case may be -- has for many generations been a basic aspiration, especially for the poorer strata of Italy's farm population with abundant supply of family labor. Aside from the social prestige involved, such possession opened the possibility of making on-farm investments to enhance output and income. However, the inheritance rules adopted from the Napoleonic Code which favor constant fragmentation of rural properties among the (until recently) numerous heirs often meant that the newly acquired status was condemned to a short duration, thus starting a new series of climbing efforts. But buying land at market prices typically in excess of its productive value means tying up savings, and/or current income for paying off indebtedness, to such an extent that not only is there no investment and working capital left but that the family subsistence level is seriously depressed (see Conclusions). The forced alternatives, renting a piece of land from year to year or making crop-share arrangements for even less than a year, were quite unsatisfactory for the aspiring peasant.

The process of spontaneous small property formation in Italy goes back to the second half of the 19th century, stimulated in part by subdivision of commons and church lands which, in the South, did not, as a rule, result in viable family-sized units. The process continued in the early 20th century in the North owing to industrial development and rising incomes; after the first world war it accelerated all over the country, aided by a number of factors -- which varied among regions in accordance with the prevailing economic structure -- such as favorable commodity prices, off-farm employment opportunities for some members of the family, war damage compensation and veterans pensions, emigrant remittances, and private and bank credit where workers cooperatives and leagues were thriving. Private intermediaries or speculators (mostly merchants) played an important part in the subdivision of medium and large properties of owners frightened by rural social unrest and constrained by tenancy protection measures. Subdivision of landed estates of charitable and public agencies also resumed.

The National Institute of Agricultural Economics (INEA) investigated the process thoroughly in the 1930s and published a series of regional monographs, as well as a final report (LORENZON) in 1938. The data show that about 500,000 peasant farmers (out of a total of 3.8 million *contadini* families) bought nearly a million hectares of farmland in the 15 years 1919-1933. About 75% of the buyers enlarged pre-existing small farms; the remaining 125,000 created new small operations, many of which were not of viable size. The research found that about 10% of the newly purchased land had to be resold during the depression years of the 1930s.

To be sure, the land market comprised many other types of operators: urban merchants and professionals, for instance, were buying family-sized farms to be co-managed and worked by *mezzadri*[4] (who, as will be illustrated later, were to become tenants and then owner-operators

---

4This is a centuries-old, semi-paternalistic system of half-share tenancy formerly typical of central Italy (notably Tuscany) which has virtually disappeared for socio-political and economic reasons and owing to over-zealous postwar regulation that had been ostensibly designed to improve the tenants' share and position. It produced some of the best family-farm managers that arose from the land reform and from the spontaneous process, including that part supported by the institution being reviewed here. There is reason to suspect that the tractor represented perhaps the *coup-de-grace* for the system.

beginning in the 1960s). Moreover, some small owners continued to round their properties even under the critical conditions of the 1930s and the war years. There is no specific information available until the early postwar years, and inter-census comparisons are problematic for definitional reasons.

In any case, the 1938 INEA report (LORENZONI) on the spontaneous inter-war movement towards peasant property, with its emphasis on motivation, different types of operators, procedures and setbacks, is very likely to have provided both information and inspiration for the land policy inaugurated by the two 1948 decrees designed to foster the further extension of the small family farm.

<u>The land market in the early postwar years.</u>

Despite their contractual weakness when facing landowners of larger dimensions and higher economic and social status, small-holders have constantly sought to expand their farms in order to gain greater economic security. The seasonal farm labor market was the last resort for one or more able-bodied *contadino* family members. The "social poverty scale" ranged from the landless or land-poor day laborer (*bracciante*) to the nearly self-sufficient small-holder.

The true, long-term cash tenant (both of family and entrepreneurial size), living on the diversified farm, was typical of the Po valley and the surrounding hillsides. *Mezzadria*, on family-sized plots often occupied by the same family for several generations, comprising typically field crops and vineyards and some livestock, with animal power, was also quite common in the Northeast, but was most typical of Central Italy (except Latium), where (in addition to including olive groves), it was usually based on centralized services and planning, mostly run by a *fattore* hired by the aristocratic or upper-bourgeois land-owner. In the South, on the other hand, there existed a variety of tenure relations, most of them implying a given division of the produce, alongside the traditional short-term rental. By definition, these tenure forms were characterized by erratic, short-term, seasonal employment.

Access to land for this assortment of actors (for whom the owning of land represented social ascent as well as economic security) depended almost solely on the availability of savings which, in turn, were a function of the vagaries of crop yields and markets, except for occasional emigrant remittances. For the *bracciante* types, there was little or no margin for savings. Other types of constraints surrounded the *mezzadri* families: the family labor-force was contractually and by necessity tied to the plot (*podere*), a serious constraint for supplementing the *mezzadro*'s share of about half the farm's net income with off-farm earnings. The *mezzadro*'s only real option for becoming a small-holder came if and when the land-owner decided to sell his estate off piecemeal, in view of their special relationship and the fact that many *mezzadri* had savings for at least a substantial down-payment.

Innovative features of the 1948 measures.

The fundamental innovations introduced by the first of the decrees -- No. 114 of February 1948, "provisions in favor of small peasant property"[5], ratified into a law by Parliament in 1950 -- were the following:

---

[5]The Italian title of the decree-law is *"Provvidenze a Favore della Piccola Proprietà Contadina"*.

o   provision of public funds for reducing interest rates on mortgages;

o   reduction, or even virtual suspension in depressed regions, of taxes and fees normally associated with land property transfer and mortgage registration; [6]

o   involvement of government officials in the certification of eligibility of applicants and in approval and monitoring of the transactions, and

o   certain restrictions on the beneficiaries in terms of the disposal of land acquired for creating or rounding family-sized farms with the benefit of public subsidies.

Mortgage loans were to be provided by agricultural credit institutions and specially authorized banks with a State subsidy on interest payments up to a maximum of 3% for 30 years[7].

The land purchases were to serve specifically one or more of the following purposes:

o   formation of new family-sized farms (particularly for landless workers),

o   conversion to fee-simple properties of parcels farmed by tenants or half-share tenants (*mezzadri*), or

o   rounding parcels of a size inadequate to support a farm family.

The same incentives were applicable to cooperatives of eligible *contadini* for buying land either for joint farming of for subsequent subdivision among members.

The decree also authorized land reclamation consortia and settlement agencies to buy land for development and subdivision among eligible farmers or their cooperatives, and to issue government-guaranteed debt instruments for this purpose, which all financial institutions were authorized to buy. These potentially important provisions were never utilized, however, because they became somewhat redundant after the second decree-law was passed one month later (see below), not to speak of the land reforms enacted in 1950.

In conclusion, the law's provisions and their practical application hinge on the individual initiative of the prospective buyer as regards the search for a suitable piece of land and the decision to avail himself of the fiscal incentives only, or of the combination of tax relief and subsidized loans. Obviously, the possibility of obtaining a concessionary loan also depended (and continues to depend) on the availability at any time of budget funds appropriated for subsidizing the interest rate.

---

[6]A substantial reduction of fiscal transaction costs has over time been extended to all family farmers. At present, the buyer is subject to the following fiscal transaction costs as proportions of the purchase price:
2% for family farmers; 10% for other farmers; 17% for all other buyers.

[7]At that time, commercial interest rates for long-term loans were around 9 percent. Subsequent legislation instead fixed the maximum rates to be paid by borrowers. For example, as will be shown later, a 1965 law (No. 590) reduced the interest rate to 1% and extended the maximum loan period to 40 years.

As we shall see in Chapter 3, this law and its later amendments have played a major role in mobilizing the market for the formation of family farms.

In March 1948, as part of a broader act -- Decree-law No. 121, ratified by Parliament as Law 159 of 5 March 1953 -- which authorized extraordinary expenditures for public works (including land reclamation and irrigation) and other development measures in the <u>southern regions</u>, creation of a special agency, the *Cassa per la formazione della piccola proprietà contadina* (Fund for the formation of small[8] peasant property) was approved. The *Cassa*'s main statutory function was to buy land and resell it -- as such or after subdivision -- to eligible individual farmers or their cooperatives. Why such a provision was not included in the earlier Decree-law (No. 114), which was entirely dedicated to the promotion of peasant property, must remain a subject for speculation.[9]

In view of the regional focus of decree 121, the *Cassa*'s operations were at first limited to the "depressed areas" (including the Central Italian region of Latium and the Tuscan Maremma, which were meanwhile added to the "South" by Decree No. 1242 of May 5, 1948). Law No. 165 of April 23, 1949 subsequently authorized the *Cassa* to cover the entire country. (That law also provided for the utilization of Marshall Plan counterpart funds for agricultural development but did not appropriate any funds specifically for the *Cassa*.)

The *Cassa*'s bylaws and regulations were promulgated by ministerial decree of September 2, 1948, at which time it became a special Office in the ministry of agriculture, administered by a board consisting of two functionaries each of that ministry and of the Treasury, under the Minister's chairmanship. (Land reclamation consortia and settlement agencies were authorized to join the *Cassa*'s board and activities, but, again, this provision was never activated.) The *Cassa*'s clientele is subject to restrictions similar to those established for beneficiaries of Decree 114, notably the impossibility for an extended period to freely alienate the property concerned, even in case of early loan repayment.

The *Cassa* and its statutes and operations are discussed in detail in Part Two of the paper.

Policy changes beginning in 1960.

The basic legislation introduced by Decree 114 of 1948 was modified somewhat in subsequent years, while new funding for its implementation was provided periodically. The decade of the 1960s witnessed an especially intensive support for family farm creation, as exemplified by the so-called First Green Plan (1965) and by Law 590 of 1965. The former introduced a new, more flexible definition of "family farmer", based on the rule of one-third as a minimum proportion of operator's and family labor input (see also footnote 25). It stressed the objective of promoting <u>efficient</u> family farms appropriate for a market economy within the framework of the newly created EEC's Common Agricultural Policy that began to be shaped in the mid-sixties. Law 590 represented a real break with

---

[8]It is not quite clear what the legislators had in mind when they qualified peasant property somewhat redundantly as "small", both in the title of the *Cassa* and in the more general legislation. In any event, the adjective was dropped from official use some years later. Perhaps the legislators simply borrowed the ambiguous notion from the 1928 legislation on long-term agricultural credit.

[9]These decree-laws were approved solely by the Council of Ministers until the election of the first postwar parliament later in 1948. Hence, there is no legislative background documentation to research.

Italy's postwar "family farm" strategy, as noted by McEntire.[10] Among other farm benefits, this law offered 40-year loans at one percent interest (which, as in the case of the *Cassa*, rose to 4 percent over the years), channeled through the agricultural banking system. Whereas previously the government had merely paid the interest rate subsidy to the banks, these loans were financed by a rotating capital fund.

The main justification for such a large public involvement is found in the explicit political objective of converting as many *mezzadri* and cash tenants as possible into owner-operators without further infringement of landowner rights (beyond the previously enacted contract and rent freezes).[11] In addition to liberal loan financing, the law also bestowed the right of preemption on tenants for the purchase of rented land, and on owner-operators for farmland adjoining their farms.

By the middle of 1975 the great importance previously attached to the formation of "peasant" property (the term is now outdated) tends to decline. The revolving fund set up by Law 590 continued to operate through the 1970s. However, under the political decentralization that began in 1970, the general program of incentives for land acquisition by family farmers became the responsibility of regional legislatures and administrations. On the other hand, the *Cassa* continues to operate as a national agency, which explains its increasing importance and impact, especially during the 1980s (see Part Two). Almost all regions have by now passed legislation providing authorization and appropriations for subsidies to family farm purchases. Special research contributed by the National association of agricultural credit institutes reveals that these regional provisions are in virtually all cases patterned after the national legislation (nearly all refer even to the basic 1928 farm credit law), and they tend to follow the national provisions regarding interest rates, with variations mostly in loan maturity (15-30 years), upper age limits for applicants and revolving capital funds versus payment of interest subsidies to lending institutions.

Beginning with Law 590, we thus find two parallel tracks: (1) the fiscal and financial incentives initiated in 1948, which were extended (with some changes) by subsequent national and regional legislation, and (2) the revolving fund created by Law 590. This fund continued to function well into the 1970s with the receipts from loan repayments. However, beginning in early to mid-1970s, the Regions became responsible for family farm-creation incentives, while the ministry of agriculture retained the responsibility for apportioning the funds among them.

Law 53 of 1956 and the First Green Plan represented a kind of bridge between the 1948 legislation and Law 590. It also introduced a novel alternative to the system of subsidizing the banks'

---

[10] "[This law] consolidated and extended the legislation in force since 1948... Government and Parliament were not prepared to follow the [1961] National Conference on Agriculture and the Rural World in offering assistance to technically advanced enterprises using wage labor. Eligibility for subsidized loans continued... to be limited to manual workers on the land, for the creation or enlargement of family farms. However... a farm could [now] depend on hired labor to the extent of two-thirds of its labor needs... This legal definition opened the way to government encouragement of farms much larger than the traditional peasant enterprise, especially with use of labor saving machinery." (McENTIRE, p. 206.)

[11] The 1966-70 Economic Program, as part of its design for supporting entrepreneurial capacity, called for the encouragement of "the transfer of ownership to those who, by means of their direct professional commitment and the contribution of their labor and capital, mean to engage in agriculture ... so as to attain wherever possible a coincidence between ownership and enterprise."

interest rates, in the form of _principal_ subsidies (averaging about 8% of the approved loan amount).[12]

---

[12]This method was most popular among existing owner-operators, half of whom opted for principal subsidy between 1961 and 1967. While trends in the average size of transactions are discussed in subsequent chapters, it is of interest to note here that the average transaction benefiting from the principal subsidy was substantially smaller (50% in the case of owner-operators) in terms of area than those where the interest rate was subsidized (as reported by MINISTERO DELL'AGRICOLTURA E DELLE FORESTE, 1968).

Chapter 3

The impact of the general financial and fiscal incentives

By 1953 the agrarian reform had expropriated or otherwise acquired more than 700,000 hectares -- three percent of Italy' total farmland -- from 3,300 private and institutional owners, under the strict deadlines established by the legislature.[13] By the end of the decade, all of this land (except for about 100,000 hectares reserved for welfare organizations, national parks and other public uses) had been reallotted to 108,000 farm families (nearly three percent of the total number), mostly in the form of what were then smallish family-sized *poderi*, but also, in areas of unusually high rural population pressure, as subsistence plots called *quote*. (The reform's beneficial effect on the farm economy and on rural employment and living conditions has lamentably been overshadowed in the public mind by the advent of the *"miracolo italiano"* in the 1960s; indeed, an entire generation of Italians is today hardly aware that there had been such a reform.)

On the whole, judging from the current agrarian structure in most of the country's productive agricultural areas, the various measures in favor of the family farm appear to have been quite successful. The agrarian reform itself helped some of the poorest rural workers to at least a piece of land for subsistence in the early 1950s. Luckily, the remaining pool of underemployed rural manpower -- especially in the South and in Sicily -- was absorbed in the next decade by the demand for unskilled labor in the northern industrial areas and in neighboring countries and by the general process of urbanization spurred by rapid economic growth in all regions. This relieved Italian public policymakers of the great predicament that plagues much of Latin America and (at least temporarily) some of the former CPEs today, namely how to insert large numbers of impoverished, under-educated rural workers productively into a modern agricultural economy and thus prevent them from swelling the ranks of the under- and unemployed in the cities.

Many owners of large and medium-sized properties in areas of social conflict were induced to sell their land by apprehension over the probability of expropriation under the impending land reform laws (which, as it turned out, retroactively froze land ownership for expropriation purposes). Apprehension continued for some time even after the 1950 reform laws were passed, at least until it became politically clear that the constitutional mandate to limit the maximum size of individual landholdings would not be translated into nationwide law.[14] Also, owners of rented farmland of whatever class (including religious and charitable institutions), unwilling to comply with the new

---

[13]Of utmost importance for the speed of the expropriation process was the provision (except in Sicily) that each expropriation decree was to be approved by an *ad hoc* joint parliamentary committee, which automatically converted the act into law and thus made it impossible for affected landowners to challenge the expropriations in court.

[14]SHEARER *et al.*, 1991 discuss the relation between agrarian reform and the mobility of land markets in Latin America. There is no doubt that Italy's agrarian reform had a profoundly positive effect on the market supply of, and demand for, farm land for family farmers:

> The fundamental obstacle to broader market-generated access to land is the existing maldistribution of land and the political power relations that flow from it. From this it follows that: (1) the institutional obstacles to broader market entry of the peasantry are symptoms of these power relations; (2) the reforms directed at the obstacles may be no more than palliatives in the absence of a prior coercive redistribution, and (3) the structural change due to market reform will be limited.

contractual rules and unable to switch to direct farm management with hired labor, were trading their land in for more secure assets.

These sales, however, did not always lead to the formation of new family farms; individual and institutional entrepreneurs, investors and speculators were also in the market. Moreover, even small owners were selling out -- often reluctantly, of course -- in order to move into other areas or activities after years of war and other limitations to mobility, and particularly between 1955 and 1965, the years of almost convulsive internal and external migration.

As regards the total volume of transactions in the immediate postwar period, available sources indicate a total turnover of about 350,000 hectares in 1948 and 332,000 hectares in 1949, i.e., less than two percent of Italy's total farmland at the time, and probably somewhat less than the prewar volume (INEA, *Annuario*, 1949). An excellent illustration of the type of operators in the land market during the same period, and of their relative importance, was provided by a (so far, unique) survey of local mortgage and transaction registers for a single year (INEA, *Annuario*, 1949). As shown in Table 3-1, all social classes, including extra-agricultural operators, had a role in the land market in 1949, either as sellers or as buyers. Land purchases by the "peasant" class (from small owner-operators to farm laborers in the table) accounted for 132,000 hectares, about 40% of the total, but the net increase in area in family farms was only 71,000 hectares because this class <u>sold</u> 61,000 hectares at the same time.[15] The survey data do not show directly what proportion of the peasant purchases went to form new family farms or to enlarge existing ones; however, it is reasonable to suppose that most of the purchases made by tenants, *mezzadri* and laborers reflected upward mobility into the class of owner-operators.

There is no doubt that, beginning in 1950, the incentives for voluntary family farm formation began to be felt in the land market. Because of time overlapping among the various programs, it was necessary to analyze separately the effects of each of the three, i.e., the fiscal and financial incentives of the 1948 decree, the innovations introduced by Law 590, and the operations of the *Cassa*.

By 1970, the provisions of DL 114 and its subsequent modifications had helped put almost 2.2 million hectares of farmland into the hands of approximately 1.1 million new and farm-expanding owner-operators (according to various issues of INEA's *Annuario*), for a modest average of two hectares per transaction. The impact was apparently smaller in the 1967-70 period, which coincided with the entry into full operation of the more favorable incentives of Law 590.

---

[15]The fact that part of the net loss of the area controlled by large and medium-sized owners who were not family farmers (the first three classes in the table) ended up also in the hands of non-farmers suggests that some of the transactions regarded parcels suitable for industrial sites or for urban development.

TABLE 3-1: Land Market Transactions by Class of Sellers and Buyers, 1949

(1,000 hectares)

Tenure Class	Purchases	Sales	Difference
Large owners	32.2	87.1	- 54.8
Medium-sized owners	44.6	66.2	- 21.6
Small non-operating owners	31.6	52.5	- 20.9
Small owner-operators	89.8	54.2	35.6
Small tenants	23.6	4.0	19.6
Mezzadri & share-croppers	15.3	1.7	13.6
Farm laborers	3.7	1.7	2.0
Industrialists	19.0	8.6	10.3
Merchants	15.3	10.6	4.7
Professionals	11.9	11.6	0.3
Other categories	45.5	34.2	11.3
TOTAL		332.4	

SOURCE: INEA, _Annuario dell'Agricoltura Italiana_, vol. III, chapter IX. The data originated in a survey by the former UNSEA (National Agricultural Statistics and Economics Office) in local mortgage and land registries.

TABLE 3-2: Number and Area of Land Transfers Under Incentives Provided by 1948
Legislation as Amended

	No. of transactions (1,000)	Area (1,000 has)	Average (hectares)
1948-60	624	1,091	1.75
1961-66	348	845	2.43
1967-70	131	229	1.75
Total 1948-70	1,103	2,165	1.96

Source: Calculated from data in INEA, *Annuario*, various years.

Not all small-farm purchasers were able to take advantage of the full set of fiscal and financial incentives. According to sources quoted by GIACOMINI, in the years 1961-67, during which the volume of transfers reached a total of approximately one million hectares, only 20-25% took advantage of subsidies on interest or principal account, i.e., the bulk of the land acquisitions were assisted only by the fiscal exemptions. The difference between the two types of beneficiaries is exemplified by the materially smaller average size (ca. 1.5 hectares) of the acquisitions with only fiscal exemptions as compared to the (also) financially subsidized purchasers (5.2 hectares). It is probable that the former were essentially laborers and small tenants who were beginning to climb the tenure ladder or small owner-operators expanding their properties, whereas the latter were already tenants or *mezzadri* who were taking over their entire *podere* from the owner[16].

After 1970, when responsibility for subsidized family farm formation was shifted to the regional governments, similar data are not available, although rough estimates of the volume of land transfers can be derived from a Bank of Italy time series of mortgage loans, to which we shall refer later.

As regards the nature of the <u>sellers</u> of land whose purchase by family farmers was subsidized in some way, a 1968 ministry of agriculture report to Parliament regarding the implementation of the two Green Plans (see GIACOMINI) relates that, while over 18% of the land was sold by family farmers, 46% consisted of smallish parcels presumably sold by various types of previously or newly urban "bourgeois" owners. Significantly, more than one-third of the land originated from medium and large private and public estates. It can be roughly estimated that about 40% of these land transfers resulted in the creation of new owner-operated farms, including some of rather modest size.

An approximate idea of the expansion of the family-sized owner-operator system through the 1950s resulting from the official land redistribution and all market transactions can also be deduced by comparing the 1961 census data with the 1947 INEA survey of types of tenure (CAMPUS 1965 and BARBERO 1967), as shown below (in thousand hectares):

---

[16]The smaller purchases may have also included a certain number of *mezzadri* who, for one or another reason, were unable or unwilling to purchase the entire *podere* as owner-operators.

- Family farm owner-operators   +1,443
- Family farm tenants   -1,581
- *Mezzadria*   -1,115
- Others   -682
- Farm enterprises managed with wage labor
  and/or share-croppers   +1,935 [17]

Interestingly, the most important increase in small owner-operated farms (two-thirds of the total) occurred in the South and Islands.

Quite exhaustive data were available (see GIACOMINI) regarding the impact of Law 590 (and its companion law, No. 817, whose resources were almost immediately allocated to the Regions) during the ten years 1966 to 1975, which comprise the bulk of the operations financed by the revolving fund (Table 3-3).

TABLE 3-3: Subsidized Loans under Laws 590 and 817

Purpose	No. approved (1,000)	Hectares (1,000)	Amount approved (bill.lire)	Average size (hectares)	Amount asked	Amount approved per hectare (1,000 lire)[1]	Ratio appr/asked (%)
Farm Creation	17.4	341.5	299,705	18.4	1,111	878	79.0
Farm Enlaragement	13.7	177.4	168,285	12.9	1,220	949	77.7
TOTAL	31.1	498.9	467,991	16.0	1,194	938	78.6

---

[1] Of 1991.

Source: GIACOMINI

These data lead to the following principal findings:

---

[17] This can be considered to some extent a statistical artifact in a comparison of the 1947 estimates with the first postwar census. The change from 1947 to 1961 reflects, for example, the abandonment of highland communal pastures and woodland that was no longer used by local peasants; the census classified as "capitalist farms", for example, large institutional properties that ceased to be operated by tenants or as commons. Moreover, the legislative restrictions imposed on non-operating landlords, as was discussed earlier, stimulated them to sell out to urban-based entrepreneurs (to the extent that the former tenants and *mezzadri* chose not to exercise preemption) or to corporate investors for both farm and non-farm use (see BARBERO 1967).

1. The bulk (about 60%) of the loans financed the formation of new family farms for over 17,000 families and covered more than two-thirds of the land area involved. The remaining transactions, of course, were for enlargement of existing farm units.

2. The average size (18 and 13 hectares, respectively) of the purchases is quite respectable, in comparison not only with the pre-Law 590 averages shown in Table 3-2, but also with the typical size of the existing, full-time family-operated farms during this period.[18]

3. The typical beneficiary of a Law-590 loan appears either to have had savings with which to supplement the loan, or to have been able to pay commercial interest on part of the price of the property: the average price per hectare approved by the provincial authority was not only about 20% below the asking price but also equivalent to only about 80% of the average price approved by the *Cassa* during the same period, and that price, in turn, was found to have been about 83% of the "asking" price during the same period. (See Part II.) Thus, if the *Cassa* price is taken as a proxy of the true market value, the average approved land price financed by the revolving fund amounted to only about two-thirds of the market price. As explained in Chapter 5, this also implies that the buyer often paid the difference in cash.[19]

4. In view of the far greater per capita demand for funds to finance the larger average size purchase, the number of beneficiaries of Law 590 was equivalent to only 3% of the number who took advantage of the older measures in the 1948-70 period.

In addition, the data show the following significant breakdown of the socio-economic status of the buyer/borrowers under the Law 590 provisions (see also Fig. 3-1):

	Number	Percent
Owner-operators	11,830	38.0
Tenants	9,332	30.0
Mezzadri	7,937	25.5
Laborers	913	3.0
Sharecroppers	498	1.5
Others	630	2.0
TOTAL	31,140	100.0

---

[18]As indicated in Chapter 2, beginning with the 1960s, we are no longer dealing basically with a welfare program for the rural poor, but with a socio-economic development strategy designed to support personal capacity and initiative for optimizing farm efficiency (to be sure, in a highly subsidized environment), within a policy of continued preferential support for the family farm.

[19]For example, let us assume that out of a total price of 100, the buyer paid 10 cash and was given a 40-year subsidized loan of 70 at 1% interest and took out a commercial bank loan for the missing 20 at 10% interest; the average interest for the borrower would have amounted to 3%. It is worth noting, in relation to the Conclusions of this paper, that a ministry circular implementing Law 590 (cited by GIACOMINI) directed the setting of a "fair" price, which was interpreted by taking into consideration the buyer/borrower's ability to pay without depressing the family's level of living, rather than relying on the prevailing "market" price.

Finally, the *Cassa* was instrumental in the transfer of 137,000 hectares to owner-operators up to 1975 (see Part II for details and later data.)

The Volume of Subsidized Long-Term Credit for Family Farm Formation.

Reconstruction of Bank of Italy records of mortgage credit transactions for family farm formation [20] by the banking system (i.e, excluding the *Cassa*) shows a rapid rise of the volume of loans (in real terms) in the first two decades (1950-70), with a sharp peak in 1966-70 under the impact of Law 590 (Figure 3-2). Subsequently, the value of such loans declined but remained relatively stable at a high level through the 1970s. Loan recoveries show a comparable, if less pronounced, trend; in recent years the gap between the value of new loans and that of repayments has tended to narrow. A similar trend appears in the area involved (Figure 3-3), estimated by applying average prices paid by the *Cassa* to loan values. (Even though, for certain periods, prices paid by the *Cassa* are by some observers considered to have been a bit generous, they tend to give, in general, a better measure of the prices of parcels suitable for family farm formation or enlargement than do the INEA averages).

It can be roughly estimated that these loans have helped finance the purchase of approximately 1.2 million hectares in 1950-91. Comparing this with the above-mentioned estimate of a total of 2.7 million hectares transacted under special provisions through 1975 confirms the finding that a large number of small transactions took place with the benefit of the fiscal exemptions only. In addition, it appears that, in terms of hectares, the *Cassa* operations after 1985 became more important relative to the volume of subsidized land transfer loans channeled through the banking system (see Fig. 3-3).

Finally, the ratio of loan recoveries to outstanding loans is an indicator of the effective average term of loans at any one time: for example, a ratio of 3% indicates an average loan term of 33 years, typical of the period of maximum Law 590 impact (40 years at 1%). Beginning in the 1970s, the ratio tends to rise gradually, reaching 7% in more recent years, which suggests a loan term of about 15 years. This trend can be interpreted as reflecting (a) the banks' reluctance to grant long-term loans during a period of high inflation, (b) increasing borrower aversion to long-term financial commitments, and (c) the legal constraints imposed on the beneficiaries of public subsidies in terms of resale and subdivision, as shown by the Cassa experience.[21]

---

[20] The series includes loans from banks' own resources with interest subsidy and those financed by revolving funds of the central and regional governments, as well as loans for the same purpose at market interest (limited essentially to a few northern regions in certain periods).

[21] This is, of course, but one type of subsidized farm credit in Italy. For example, as of 1991, the journal *Terra e Vita* (Vol. 32, No. 31/32, 3-23 August 1991), lists several other sources of officially subsidized credit through the banking system for farmers in general, some with additional subsidy for family farms, to wit:

Working capital loans for 6 months at 8.25% interest.

Various schemes for loans of 2 to 5 years for farms that had serious crop damage from adverse weather conditions (hail, etc.), at 5% interest for all, 2.85% for family farms (except for one type of loans where all farms pay 2.85%).

Farm machinery purchase loans for 5 years at 5.7% for all farms.

10-year loans for irrigation and greenhouse equipment at 6.1% for all.

(continued...)

Chapter 4

Land prices, farm wages and general inflation: peasant purchasing power in terms of land

Even after the pressure of landless and land-poor peasantry on the country's very limited arable land resources eased in the 1960s, a number of factors combined to cause a steady rise in the real price of that type of land through the 1970s (Fig. 4-1); demand from young entrepreneurial farmers who had reached an economic limit of productivity per hectare and needed to enlarge their holdings to keep their incomes in line with the rapid growth of non-agricultural incomes; abandonment of farmland that could not be mechanized; encroachment of urbanization on good arable land everywhere; the latter-day reluctance of owners of land who had abandoned farming to divest themselves of the land, and finally, the ineluctable expectation of inflation. It should be noted that land values in Italy at the end of the war already stood at relatively high levels "owing to the great imbalance between demand and supply in the past". (Barbero 1964, pp. 382-383.)

Comparisons of three main variables can be used to indicate the relative capacity of Italy's "peasantry" to buy land for the formation of new family farms or for rounding existing ones over a period of some 30 years beginning in the late 1950s, when there was still much rural underemployment and poverty (despite the land redistribution and attempts to make various tenancy

---

[21](...continued)

15-year loans for farm dwelling construction or improvement, for family farms at interest of 4.5% to 7.5% according to location, and for others from 7.5% to 10.4%.

All-purpose farm improvement loans, including refinancing, for all farms, cooperatives and agro-industries (but also including land purchases for the formation of "peasant property"), "indicatively" for 10 years and at variable interest, "currently" 11.7%.

Finally, there were several agro-industry-type loan programs for 10 to 20-year loans at interest rates ranging from 4.55% to 8.35% according to location (the South and mountain areas pay the least interest, depressed areas of north and central Italy pay 6.1%).

The Treasury's "reference rate" (TR) is the basis for the determination of subsidized interest rates to borrowers. It is announced every two months and it ostensibly covers the financial cost of money for the credit institution plus a commission charge. This represents the maximum rate that can be charged to borrowers. The minimum is fixed as a percentage of the TR; at present it is 30% for depressed areas and 60% for the rest of the country. This means that payment of treasury subsidies are limited to 70% of the TR in depressed areas and 40% elsewhere. Regional laws may supersede these provisions if the subsidies are paid from regional funds. (Note, however, that it is reported that one southern Region as of 1992 owed one of the principal southern banks 200 billion lire - about US$167 million - of reimbursements for subsidized interest on officially approved loans; this, of course, raises the bank's transaction cost and is thus a potential deterrent for bank participation in such schemes.)

Bank of Italy data show that government budget outlays for all "interest-reducing" subsidies in agricultural lending rose from an annual average of 1.7 billion current lire in the 1950s to 16.6 billion in 1960-65 and to 53 billion in 1966-70, while the proportion that went for land purchases declined from about one-fourth to around 10%. At the same time, however, appropriations to revolving funds for land purchases rose to 250 billion current lire for the five years 1966-70, when they accounted for more than one-half of all appropriations to revolving funds for agriculture (PONTOLILLO, Tables 3 and 4).

arrangements more equitable): between the trends of land prices and farm wages and of both with the trend of general inflation (as measured by the official cost-of-living index - CoL). (Tables 4-1 to 4-3 and Figs. 4-1 and 4-2).[22]

In Emilia-Romagna, land prices declined against the CoL between 1957 and 1966, after which they began to exceed the CoL substantially through the early 1980s, only to stabilize, both *per se* and against the CoL. On the other hand, farm wages (with the support of well-organized and politically influential unions and strong demand for industrial labor) began to outrun land prices as early as 1963. Wages had risen more than three times as fast as land prices from 1957 to 1984.[23]

In Tuscany, on the other hand, land prices lagged behind the CoL through 1972, as a reflection of the large-scale abandonment of hill-land farms in those years and the *mezzadria* crisis which induced many traditional landowners to sell. Land prices rose faster than the CoL from 1975 to 1981, then began to lag again and finally pulled even in 1990. But here, too, farm wages outran the CoL as early as 1963, stabilizing at a level of about three times the CoL index for the past ten years. Farm wages had begun to exceed land prices substantially in the early sixties, reaching a level of three to four times the index of land prices since the mid-1980s despite the strong residential demand for rural properties.

In Apulia -- where rural unemployment and poverty were endemic well into the 1950s and which is still relatively underdeveloped -- land prices have lagged significantly behind the CoL until 1966 and again from 1984 onward; the trends were virtually parallel in the intervening years. Here, too, strong labor unions and industrial labor demand from the North and from west-central Europe pushed farm wages greatly ahead of land prices as early as 1963; the ratio peaked at about 4:1 in the late 70s, and has remained there.

---

[22]The data for land prices and farm wages refer to specific regions (or provinces), where the *Cassa* has been most active; as will be seen, there are some significant differences (in trends as well as absolutes) among the three regions, located, respectively, in the North (eastern Po Valley), Center and Southeast. National averages would be meaningless.

[23]ROSATO (1991) analyzed the factors determining changes in farmland prices in neighboring Veneto region from 1960 to 1988. While most of his interesting findings lie beyond the scope of the present study, a few observations are worth reporting here:

o   real land price changes in the central plains and piedmont were on the order of 300%, compared with 100% and 150%, respectively, in the mountains and the peripheral plains;

o   changes in efficiency -- as expressed in a tripling of farm acreage per active person in agriculture -- rather than changes in product prices, have been one of the determinants of changes in land prices;

o   urban land per capita has doubled, while farmland has shrunk in the absolute;

o   alternative investments in urban real estate and securities have helped to steady farmland prices;

A sample analysis of farm sales in 1986-1988, in addition, showed that

o   25% of the sellers were family farmers, and 43% of the buyers.

TABLE 4-1: Land Prices per Hectare (LP/ha), Farm Wages[1] per Year (FW/yr)
and Years of Wages Required to Buy One Hectare (LP/FW),
in Selected Areas and for Selected Years, 1957-1990.

(Wages and prices in thousand current lire)

| Year | Emilia-Romagna | | | Tuscany | | | Apulia | | |
	LP/ha	FW/yr[2]	LP/FW	LP/ha	FW/yr[3]	LP/FW	LP/ha	FW/yr[4]	LP/FW
1957	630	403	1.56	369	375[a]	0.90	552	283	1.95
1960	604	406	1.49	351	378	0.93	526	276	1.91
1963	594	698	0.85	346	623	0.56	539	482	1.11
1966	689	1038	0.67	334	866	0.39	564	724	0.78
1969	1266	1106	1.14	504	996	0.51	754	834	0.90
1972	1333	1269	1.05	594	1012	0.59	834	1106	0.75
1975	2412	2163	1.12	1206	2427	0.50	1512	2196	0.69
1978	4426	4349	1.02	1779	3864	0.46	2241	4681	0.48
1981	7384	8202	0.90	3361	7499	0.45	3743	8190	0.46
1984	6230	13298	0.47	3481	12266	0.28	3650	11977	0.30
1987	7824	14984	0.52	3343	14977	0.22	4407	n.a.	---
1990	10007	17341	0.58	5020	17566	0.29	5314	17471	0.30

[a] Extrapolated

[1] Minimum gross total annual remuneration of permanent or indefinite contract farm workers.
[2] Province of Ferrara (except Ravenna for 1966 and 1978).
[3] Province of Siena, except Grosseto for 1969, 1981 and 1987, and Arezzo for 1957 and 1972.
[4] Province of Foggia, except Bari for 1984.

SOURCES: For Wages: ISTAT, _Annuario Statistico Italiano_ for selected years. - For Land Prices: INEA, _Annuario_ for selected years.

TABLE 4-2: Indexes of Average National Cost of Living (CoL), Compared with Indexes of Land Prices and Farm Wages in Selected Areas, for Selected Years, 1957-1990 (1957=100).

Year	CoL	Emilia-Romagna		Tuscany		Apulia	
		Land	Wages	Land	Wages	Land	Wages
1957	100	100	100	100	100	100	100
1960	107	96	101	95	101	95	98
1963	125	94	173	94	166	97	170
1966	140	109	257	91	231	102	257
1969	149	201	274	137	266	137	295
1972	174	212	315	161	270	151	391
1975	269	383	537	327	647	274	776
1978	416	703	1,079	482	1,030	406	1,654
1981	692	1,172	2,035	911	2,000	678	2,894
1984	1,023	989	3,300	943	3,271	661	4,232
1987	1,233	1,242	3,716	1,041	3,994	798	n.a.
1990	1,464	1,588	4,303	1,360	4,684	963	6,173

SOURCE of CoL: ISTAT, _Annuario Statistico Italiano_. For sources of land prices and wages, and for notes: See Table 4-1.

TABLE 4-3: Ratio of Indexes of Farm Wages in Selected Areas to Indexes of National
Average Cost of Living (1957=100) for selected Years, 1957-1990.

YEAR	EMILIA-ROMAGNA	TUSCANY	APULIA
1957	1.00	1.00	1.00
1960	0.94	0.93	0.92
1963	1.38	1.32	1.36
1966	1.84	1.64	1.84
1969	1.84	1.77	1.98
1972	1.98	1.54	2.25
1975	2.00	2.39	2.88
1978	2.59	2.46	3.98
1981	2.94	2.87	4.18
1984	3.23	3.17	4.14
1987	3.01	3.21	n.a.
1990	2.94	3.17	4.22

Source: Table 4-1.

The bottom line in the context of the present study is perhaps the trend in the purchasing power of farm wages in terms of farmland (Table 4-1 and Fig. 4-1), which rose dramatically from the late 50s in all three regions. In Tuscany and Apulia, one hectare of average land could be bought with less than one-third year's labor by 1984 (and with half a year's labor even in Emilia-Romagna), compared with 1½ years' labor in the late 1950s in the latter, and two years in Apulia (but only one year in Tuscany). The considerable appreciation of rural wages in terms of the CoL reinforces the impression of greatly enhanced peasant purchasing power because it has engendered constantly increasing growth in discretionary expenditures and savings. (Remittances of savings from out-of-region employment obviously also played an important part.)[24]

---

[24]Note, however, that the land prices are regional averages. As Rosato points out (see previous footnote), prices for good land rose two to three times as fast as those of marginal land. Thus, the purchasing power of farm wages in terms of good land rose less dramatically than the average indicates.

Part Two: The Cassa Per La Formazione Della Proprieta Contadina

Chapter 5

History, structure and functions of the *Cassa*

One-and-a-half years before the enactment of Italy's first land reform law (for the southern region of Calabria), and hard on the heels of Legislative Decree No. 114 that provided general incentives for purchasers of family farms (see Chapter 2), a Fund for the Formation of Small Peasant Property -- the *Cassa* -- was created by one brief article of a sweeping Legislative Decree, No. 121 of March 5, 1948. This decree was the first installment of the special measures establishing incentives, massive public investments and new institutions for the development of the depressed South and the two island regions, preceding the election of Italy's first postwar parliament. (It was ratified by Parliament in 1953.)

Forty-four years later, Presidential Decree No. 1168, issued in April 1983, reformed the *Cassa*'s bylaws substantially, giving it an indefinite institutional existence, together with an autonomous administrative structure. Previously, a 1978 Presidential Decree had confirmed the *Cassa*'s usefulness and indefinite life (along with a number of other public agencies). The president and vice-president are appointed by the minister of agriculture. The 14-member board of directors, who serve four-year, non-renewable terms, is composed of the president and vice president, two representatives of the ministry of agriculture, one from the budget ministry and two from regional administrations; one member represents *Cassa* personnel. The remaining six members represent farm operator associations, including two from the cooperative movement.

An executive committee of five members, which includes two farmer representatives, is empowered to make specified operational decisions, in addition to those specifically delegated by the board. A board of financial controllers, named by the minister of agriculture, is composed of officers of the agriculture and treasury ministries and the Court of Accounts.

The 1983 decree did not introduce any substantial novelties with respect to the acquisition of funds by the *Cassa*: funding was to continue to come basically from the central government budget, in addition to any income from the *Cassa*'s operations, program funding from other institutions, "contributions from public administrations, associations, and public and private agencies", expense reimbursements, and "other possible revenue". The decree does not refer to the provisions of DL 114, discussed in Chapter 2 (which had remained unimplemented), authorizing certain land development and settlement agencies to issue debt instruments for farm subdivisions.

Two important innovations to the *Cassa*'s statutes were introduced by Ministerial Decree of 8 July, 1991: it may "participate ... in initiatives in the sphere of cooperation with developing countries" (Article 2), and it may "supplement its own financial resources ... through recourse to the internal and external [capital] market ..." (Article 3 (c).

The *Cassa*'s internal organization is relatively simple: it consists of only four divisions, whose directors are responsible to the Director General; he, in turn, answers to the President. Key staffing aspects, according to the 1990-approved table of organization, are indicated, as follows:

General affairs, personnel and coordination of interregional offices.
Total staff: 33

Research, programming, data processing and technical assistance to borrowers. Staff consists of only four senior officers, with 11 assistants and operators. (As indicated elsewhere, agricultural extension and accounting services are contracted directly by the farmers with local NGOs or consulting firms.)

Technical services. The main center of action that deals with land purchase and resale, on-farm investments and loan guarantees. It is staffed by 11 senior officers, 14 "professional technical collaborators" and 16 assistants and operators.

Administration, asset management and accounting (including relations with borrowers, budget and financial control and management of repossessed land). It has a staff of 27, of whom nine are senior officers.

The total number of approved positions (supplemented by a certain number of temporary contract personnel) is 120, of which four are earmarked for senior managers and 30 for senior operating staff.

Notable -- although in common with most public agencies in Italy -- is the lack of any kind of monitoring and evaluation unit; the list of functions assigned to the Programming, research etc. division by the 1990 regulations makes no mention of such a function. In fact, no systematic evaluation of the organization's functions, procedures and impact has ever been attempted.

The *Cassa*'s operational budget -- 7.6 billion lire (US$ 6.3 million) in 1991, net of taxes which amounted to double that sum -- represented only 6% of its capital expenditures. Since the *Cassa* became autonomous in 1983, this proportion has fluctuated annually between 3% (1988) and 11% (1985). (See also Chapter 6.)

In addition to being authorized to acquire farmland, including fixed assets, by purchase, donation or exchange, and to develop and resell it to individual family farmers as well as *pro indiviso* to cooperatives, the *Cassa* may guarantee loans, provide legal and technical assistance to the buyers and disburse loans as well as standard subsidies for land improvements to their mortgagees, and plan and implement parcel consolidations. Setting of resale prices and of interest rates and terms of loans is the responsibility of the board of directors (i.e., these criteria are, wisely, not fixed rigidly by statute).

Following the creation of the Regional Development Agencies (*Enti Regionali di Sviluppo*) in the early 1960s (which, in the regions concerned, also took over the functions and personnel of the seven former regional land reform agencies, and which were absorbed by the regional governments in 1976), the *Cassa* was directed to channel some of its funds through these agencies as well as to utilize their technical services.

Among the most important norms of the *Cassa*'s *modus operandi* are the following:

Normally, the *Cassa* requires both an offer to sell and offers to buy before it embarks on any project, meaning that seller and buyer(s) must have some kind of informal understanding before the

Cassa is approached. *Cassa* personnel sometimes act informally in helping to establish the connection.

Offers to sell are examined in the first instance by a committee of three, representing, respectively, the *Cassa* and the provincial and regional authorities concerned. Their recommendations regarding the acquisition price and conditions are then formally approved by the regional government and reviewed by an ad hoc "technical" subcommittee of the *Cassa*'s board, which makes its recommendations to the full board.

A number of regulations are clearly designed to avoid abuse for speculative purposes of the highly subsidized system as well as outright fraud; in addition, some of them still reflect the early postwar policy concern with keeping people on the land, and, in the absence of general legislation, avoiding farm fragmentation:

1. Each applicant for a *Cassa* loan must be certified by the regional agricultural authorities as a *bona fide* family farmer (the legal term is *coltivatore diretto*), i.e., a full-time, resident farm operator.[25]

2. The buyer/borrower may not have sold or otherwise alienated to others any farmland during the preceding two years except with official, local authorization.

3. A five-year trial period for purchasers of land from the *Cassa* was required by Law 590 of 1965 which contains detailed "Measures for the Development of Family Farm Ownership". This

---

[25]The criteria governing the definition, for entitlement purposes, of a "family farmer" and a "family sized farm" have, of course, changed substantially over the years, as farm policy responded to the rapid decline of the farm labor force and to Italy's accession to the European Common Market. Wisely, they have always been relatively flexible. Until 1954, the 1928 concept of "small peasant property" for entitlement to subsidized farm improvement loans was used: the total area "that can normally be farmed by the owner's family and which shall not be smaller than the minimum area necessary for the rational management of a family farm". Subsequently, in response to two new laws, and on the basis of Article 846 of the Civil Code, the Ministry of Agriculture issued an administrative rule designed to assure that a small peasant property "guarantee a minimum of self-sufficiency" for the purchaser, as well as "the absorption of the ... family's labor". As early as 1961, a national rural life and agriculture conference amplified the concept of "family farm enterprise" in terms of "adequate dimensions for the life and labor of a normal family, so that it allows autonomy of management <u>and sometimes even the development of superior forms [sic] ... which, together with the family labor, also admit the use of outside salaried work force</u>". The first Green Plan law, also of 1961, introduced into the general eligibility criteria for being considered a *bona fide* "*coltivatore diretto*" the so-called "one-third rule" (first devised in 1949 in connection with entitlement to automatic farm lease extensions), i.e., that, aside from requiring that the applicant be a resident operator of the farm in question, <u>the total family labor force shall not be less than one-third of the total labor requirement for normal farming operations</u>. (See GIACOMINI, pp. 356-360.)

Law 590 of 1965 contains detailed "Measures for the Development of Family Farm Ownership" which, in part, also reflect the dramatic changes that had occurred in the country's agrarian structure since the end of World War II. In its reprint of pertinent legislation (*Principali Leggi* ... 1990), the *Cassa* appends the following footnote to this law:
The criterion of self-sufficiency on which the previous concept of "small peasant property" was based has been made obsolete by the new concept of farm efficiency, which combines entrepreneurial capacity and the economic size of the farms so as to make it possible for the latter to participate effectively in today's market dynamics.

This reflects the radical change in the socio-economic context that took place in the early '60s; leading farm-supply firms, as well as farmer organizations, were pressing for an explicit policy of agricultural modernization aiming at greater efficiency, higher labor productivity and market orientation, leading to farm size enlargement and increasing producer cooperation and, later, to a great expansion of part-time farming, particularly in the center-north.

was increased to 10 years by Law 817 regarding financial appropriations. The reformed trial period of ten years is the same as the minimum age of the loan at which it can be repaid in advance. (Following expiration of the trial period, the buyer may also apply for transfer of ownership to a cooperative that he may have joined.)

If a buyer withdraws within this period, he must repay all the tax exemptions (see Chapter 2), except in the case of *force majeure*. In case of withdrawal, he also loses the right to compensation for land improvements that were directly subsidized. The provincial office of the ministry of agriculture, originally, and currently the regional administration, certifies the nature of the termination.[26] It is also required to initiate revocation, during the trial period, of the original certification in case the buyer is found not to operate the farm himself.

4. Finally, the law places a 30-year restriction on the property concerned that prevents its subdivision in case of alienation. This restriction -- which is similar to that governing agrarian reform allotments -- remains in force even in case of early repayment of the loan, except in case of succession.

Interest rates on *Cassa* mortgage loans - as noted earlier - are set by the board of directors, although in practice they have been kept roughly in line with the concessionary, government-subsidized loans for small property formation and for on-farm fixed investments (at their minimum levels) by the agricultural banking system (see Chapter 3). From the creation of the *Cassa* mechanism in 1948 until about 1955, the loans carried an interest of four percent. From 1956 to 1972, it was one percent. Long-term bank mortgage loans were paying a market rate of around 8 to 9 percent until about 1969. By 1975 commercial mortgage interest had risen to about 14 percent: the *Cassa*'s rate was doubled, to two percent, in 1973. Commercial rates, and inflation, showed no signs of abating, so that the *Cassa* board doubled the concessionary rate once more, to four percent, in 1978. From then until 1992 this rate has represented at least a 10 percentage-point concession below the going commercial rate, and even more in 1981-85 (Fig. 5-1).[27] (The implications of this subsidy are discussed in Chapters 7, 8 and 9.)

Standard transaction costs (which reflect, of course, the special tax and fee relief mentioned earlier and also include a commitment fee), averaging about 3.5% of the price of the land, are added to the loan principal.

In principle, the *Cassa* gives two years of grace for overdue payments but charges 10% interest on the overdue balance. After that, it could proceed rapidly to foreclosure under existing Italian law

---

[26] It also certifies that the grantee is a "manual worker of the land" (*lavoratore manuale della terra*). Though somewhat outmoded by mechanization and by the changes in other criteria, as discussed in the previous footnote, this wording is still in effect and is now very liberally interpreted.

[27] Recently, the *Cassa* board has been empowered to authorize, on demand from borrowers in mountainous or otherwise "disadvantaged" areas (CEC definition), to reduce the rate to two percent even for outstanding loans. This is a response to increasing difficulties for marginal farmers to keep their payments up in the face of declining farm prices.

and procedures. In fact, until recently[28] borrower delinquency or default has not been a significant problem, thanks to the self-starting selection process, the thorough investigative process and, last but not least, the highly negative real interest rate on the loans.

In the mid-1970s the board decided that the minimum land purchase must be equivalent to at least 30 million lire of gross value of product at 75/76 prices (except for *mezzadria* conversion, for which the minimum is 15 million lire). This is equivalent to about 20 hectares of good land; the limit is reflected in the average size of the farms sold by the *Cassa* from 1970 (see Table 6-2).

Because 30-year mortgage loans are supposed to cover only land and fixed investments, the *Cassa* does not officially purchase operating capital existing on the farm. This must be, and normally is, negotiated directly between seller and buyer and, if necessary, financed separately through commercial, including subsidized, credit operations.

Despite the existence of four small field offices (soon to be expanded to five), and the cooperation of regional government agencies, the *Cassa* has remained a highly centralized agency. Non-financial functions authorized by the law, such as technical assistance and direct investment in farm improvements, have been performed only exceptionally. Loans for on-farm fixed investments are made occasionally, at rates set by the board, but not less than 4%; the *Cassa*'s clients typically turn to other public programs providing direct subsidies or subsidized credit, or to the credit market, for this purpose.[29] No parcel consolidation projects have been undertaken, even though (or perhaps, because) Italy has no specialized service for this purpose.

The *Cassa* uses an *ad hoc* version of standard Italian farm appraisal concepts for its technical and financial appraisal of properties offered for sale. One concept is "total value of gross production" (PLT), estimated from known local parameters plus a combination of observable phenomena -- such as topography, soil quality and the prevailing cropping pattern -- as a proxy of salable farm output. The other is "net returns to land" (BF). In the *Cassa*'s practice (as well as for appraisal of subsidized farm purchase loans through the banking system), this is based on a more-or-less arbitrary ratio to PLT that varies essentially in accordance with the degree of fixed investment. Sample data from the *Cassa*'s files (see below) show that BF has fluctuated around one-fifth of PLT for the typically diversified farms in the northern and central regions and around one-fourth in Sicily, where the properties have frequently comprised citrus plantations.[30] A standard rate of capitalization is applied to BF to arrive at a first approximation to a fair market price to be negotiated with the seller. (The

---

[28]Beginning with the 1988/89 crop year, according to *Cassa* management, as many as one-third of the current borrowers (mostly grain farmers and other single-croppers on marginal land and stock raisers above 600 meters altitude but also some of the very intensively managed enterprises) were having difficulties with their loan amortizations. This is said to be related to declining real prices for many farm products (see Fig. 4-1).

[29]In 1991, for example, the *Cassa* made only four farm improvement loans for a total of 2.36 billion lire (of which 455 million for one cooperative). However, the *Cassa* guaranteed 40 concessionary loans from other sources for its clients, for a total value of 5.9 billion lire.

[30]The state of the case files for years prior to 1971 did not allow analysis of all variables within the time allotted to the research. Since Total Gross Product (PLT) and Returns to Land (BF) were not retrievable in about half the cases in the sample, the analysis of these variables concerns only cases initiated in 1971 or later. In any case, the percentage calculations -- frequency distributions as well as others -- are based solely on the respective total number with information.

nominal rate varies little between regions and over time: in nearly all cases of the sample studied, it fell between four and five percent and the overall mean is 4.6% with a standard deviation of .73. The lowest means are found in the North (4.1 and 4.3%, respectively); in Tuscany the rate rises to 4.7%, and in the two southern regions and Sicily it averages close to 5%.)

A regionally random 10% sample of "offers for sale" was drawn from more than 10,000 case files of applicants for land purchases in the six regions of Italy where the *Cassa* has been most active since 1948[31]. Only 53% of the 1,040 applications examined reached the contract stage: 37% were withdrawn before they reached the *Cassa*'s Board; the remaining 10% were either rejected by the Board[32], or they encountered subsequent problems. Lack of funding was adduced in only a few cases and in particular periods. The high proportion of dropouts may be in part related to the fact that only 14% of the cases that were not withdrawn reached the Board in less than six months from the date of application; 28% took from seven to 12 months, and another 19% more than two years. Most of the cases (39%) remained in the pipeline for one to two years. Following Board approval, the process gains speed: contracts were signed within six months in nearly half the cases, and another one-fourth within the year. According to *Cassa* management, the delays tend to be due to temporary shortages of loan funding.

More than three-fourths of the applications for purchase were from single buyers. More than half of the purchases were designed to expand an existing farm, reflecting the coincidence of the reorientation of public policy from welfare to efficiency at the time the *Cassa*'s funding was multiplied; the remaining cases were almost evenly divided between creation of new farms, moves to a new area and - probably more typical of the early years - subdivision of the purchased parcel into smaller units.

For the sample as a whole, and over the years, the average purchase price appraised by the *Cassa*-led commissions has been about 25% below the average price demanded by the seller.[33] On the other hand, the Board seems to have been in almost all cases satisfied with the commissions' appraisal: the average Board-approved price is only four percent less than the appraisal average. For the record, the average price paid per hectare over the years (in terms of 1991 lire), including buildings and other fixed investments, was 15,267,000 lire, equivalent to about US$12,700 (at the approximate market exchange rate of 1,200 lire per dollar).

For the three major regions taken together, the average Board-approved price per hectare (in 1991 lire) was equivalent to four times the estimated (pre-transaction) gross value of production; the ratio was 1:4.09 for the northern and central regions and 1:3.77 for the South. At the respective ratios of estimated returns to land to gross product used for loan appraisal shown above, estimated pre-transaction returns to land have been equivalent to between five and six percent of the price paid

---

[31] Lombardy and Emilia-Romagna (North); Tuscany (Center); Apulia-Basilicata and Sicily (South).

[32]The most frequent reason for rejection has been the finding that the parcel concerned could not meet the economic viability criterion in terms of its potential for producing an acceptable family income plus the loan amortization. The next most frequent cause of rejection was "failure to submit the requisite documentation".

[33]There is a strong suspicion that this almost standard difference between offering and appraisal reflects in reality prior (and tacitly tolerated) understanding between buyer and seller to remain close to the asking price, with the difference payable in cash or through informal deferred arrangements between the parties.

for the property. While these ratios are not necessarily based on analytical calculations, this appears to justify a subsidized interest rate (see Chapter 8). GIACOMINI (p. 371) cites Ferro's[34] opinion that this standard procedure is but a way to "make the incidence of returns to land coincide with that of the loan amortization rate", i.e., the predominant criterion is repayment capacity, rather than family income, a concept that is not even used in the loan appraisal.

Completed transactions were examined for 114 cases, covering more than 15,000 hectares, in favor of farming cooperatives which comprised nearly 20,000 members.[35] This represents one-half of the total land sales to cooperatives, and 70% of the number of beneficiary cooperatives, through 1991 (see Table 6.2). (Note, however, that only about 1,100 cooperative member families can be assumed to have been directly affected; the number of 20,000 includes retired or otherwise inactive members as well as multiple family memberships). The average transaction comprised 132 hectares. The transactions concerned almost entirely supplementary land for existing cooperatives in the North; only five new ones were created -- two in the North and three in the South. The vast majority of the farming coops are found in the northern region of Emilia-Romagna, the cradle of (voluntary) collective farming in Italy.

According to *Cassa* estimates, approximately 5% of the land transfers it has financed through 1993 -- and an even greater proportion since 1991 -- represented intra-farm family transactions from owner-operators of pensionable age (or with disabilities) to junior members. In these cases, the *Cassa* can play an important role by paying the retiring owner-operator for his/her stake, out of which he/she can then compensate the non-farming potential co-inheritors, at the same time stipulating the usual 30-year loan agreement at subsidized interest with the new owner. In fact, the *Cassa*'s most recent regulations allow it to mediate such intra-family transactions as soon as the incumbent owner-operator reaches age 60, or has a certified disability, or has "commitments in other sectors of production".

The objective of the European Council's Regulation 2079/92 (published in Italy's *Gazzetta Ufficiale* of 17 September 1992) is the encouragement of older farm-operators to retire early in favor of the younger generation, regardless of whether or not the successor is a family member. For this purpose, the minimum retirement age is reduced to 55 years and funds for special pension supplements are made available to the member countries. The funds are used to pay the early retiree, until he/she reaches the member country's statutory retirement age (and provided he/she has been farming for at least ten years), an annuity or to bring the total annuity of previously pensioned farmers up to the above maxima.

A number of conditions for specific country implementation outlined by the new Regulation in its Article 6 are recognized to be of potential importance for the future activities of the *Cassa* (and of the regional administrations concerned) to wit:

---

[34] O. Ferro, *"Attuali problemi ed aspetti del mercato fondiario italiano"*, in *Agricoltura delle Venezie*, No. 5, 1967.

[35] The *Cassa* participates in management of all coops throughout the life of the loan (and sometimes beyond), and a representative of the *Cassa* is required to preside over their boards of controllers. This is reportedly welcomed by the cooperatives. Though many problems are encountered in the newly formed cooperatives, the *Cassa* tends to help solve them early on by quietly promoting replacement of problem members.

o the farm released by the pensioner must be of economically efficient size in relation to the capacity of the successor; (the *Cassa*'s most recent criteria define economic viability of a farm its capacity to produce a total value of salable production of not less than 60 million lire - ca. US$37,500 - except that in specific cases this limit can be halved);

o the land released to the successor must be farmed for at least five years and in accordance with environmental requirements;

o the land released may be included in parcel consolidation operations or simply in parcel exchanges.

The perhaps most relevant section of this article reads as follows (our translation and underlining):

<u>Furthermore, the member States may provide that the released lands be taken over by an organization that commits itself to turn them over subsequently to purchasers who meet the conditions of the present regulation.</u>

Preliminary proposals contemplate a possible collaboration between the *Cassa* and the farm social security service (SCAU). According to SCAU estimates, there may be as many as 35,000 farms comprising 385,000 hectares potentially interested in *Cassa* mediation. This would be another instance of the institution adapting its objectives and parameters to changing rural conditions.

The *Cassa's* resources and impact; The cost of the subsidies

Except for a three-billion current lire (43.6 billion 1991 lire), 37-year loan from the Treasury's *Cassa Depositi e Prestiti* (CDP) in 1960 (which will be paid off in 1997), the *Cassa* has not resorted to borrowing. The CDP also transferred, between 1952 and 1977, a total of about 1,000 billion (1991) lire of its earnings to the *Cassa*'s capital fund. From 1948 to 1991, including the cash transfers from the CDP, the *Cassa* received 3,236 billion (1991) lire from the government budget by means of *ad hoc* legislation, equivalent to about US$2.7 billion[36]. (Table 6-1.)

TABLE 6-1: *CASSA* Capital Resources and Land Purchases, 1948-1991, By Periods.

Period	Capital Replenishments		Land Purchases	
Total	Annual Average	Total	Annual Average	
	(Million 1991 lire)		(hectares)	
1948-1955	84,272	10,534	15,858	1,982
1956-1960	345,995	69,199	35,294	7,059
1961-1965	141,423	28,285	23,289	4,659
1966-1970	332,185	66,437	27,891	5,578
1971-1975	152,650	30,530	34,221	6,844
1976-1980	105,779	21,156	32,368	6,474
1981-1985	185,000	37,000	32,155	6,431
1986-1991	442,800	88,560	76,811	12,802
TOTAL	3,236,463	73,556	284,731	6,352

Source: Calculated from data furnished by *Cassa*.

According to *Cassa* management, the rhythm of land purchases and resales is entirely a function of the periodic additions to its resources. The low demand for the *Cassa*'s intervention in the initial eight years can be attributed to the implementation of the agrarian reform. There has been little

---

[36]There was an allotment of about one billion lire from ERP (Marshall Plan) counterpart funds in 1949 (equivalent to about US$1.5 million at the time), which is probably comprised in the two billion lire budget appropriation of that year.

variation in the annual averages of farmland purchased by the *Cassa* through the 1981-1985 period. *Cassa* management asserts that the doubling of activity in the most recent five years was a reflection of a sharp increase in market demand due to high farm incomes and to the termination of many farm leases under a new law, as well as to the more effective pressure for additional funds by the newly autonomous *Cassa*.

The *Cassa*'s current expenditures are normally funded from interest payments by borrowers; indeed, in some years surpluses of interest receipts have reportedly been transferred to the capital account.

The *Cassa*'s net assets at the end of 1991 were valued at 1,090 billion lire, of which 964 billion (US$800 million) represented the agency's portfolio of farm mortgage loans. As discussed in Chapter 5, borrower defaults have thus far been negligible (in any case, they are nearly always covered by the quick reallotment to new buyers of repossessed farms). Thus, a 70% capital erosion is indicated solely from inflation, i.e., without considering the opportunity cost of the moneys provided at market rates of interest (see below).

Period	To Family Farmers					To Farming Coops			
	Number		Hectares			Number		Hectares	
	Total	Annual Average	Total	Annual Average	Av. p. sale	Total	Annual Average	Total	Annual Average
1 9 4 8 - 1956	7351	817	18875	2097	2.6	[a]		[a]	
1 9 5 7 - 1970	5578	398	66991	4785	12.0	[b]58	2.5	[b]15361	668
1 9 7 1 - 1980	2243	224	65862	6586	29.0	20	2	1838	184
1 9 8 1 - 1991	4086	371	101414	9219	24.8	88	8	11995	1090
TOTAL	19258		253142	--	--	166	--	29194	--

[a] Not available. [b] 1948-1970.

Source: Calculated from data furnished by *Cassa*.

In its 42 years of existence, i.e., through 1991, the *Cassa* acquired a total of nearly 280,000 hectares, of which 253,000 were resold to about 19,000 individual farmers, and 29,000 to 166 production cooperatives[37] (Table 6-2). This is equivalent to about 40 percent of the area redistributed by the agrarian reform in the 1950s and about one-fifth of the total estimated subsidized land market transfers to family farmers during the past 40 years (see also Fig. 3-3).

---

[37]Data regarding the number of <u>active</u> members of these cooperatives at the time of purchase are not available.

<u>TABLE: 6-3:</u> Average Prices Paid for Farmland (including fixed investments) by *Cassa* in Selected Years and Selected Areas and Nationwide, and Respective Indexes, 1958-1990.

(Million current lire)

Year	Emilia-Romagna		Tuscany		Apulia		All Regions	
	Lire	Index	Lire	Index	Lire	Index	Lire	Index
1958	0.583	100	0.413	100	0.219	100	0.298	100
1960	0.460	79	0.229[a]	55	0.289	132	0.285	96
1963	0.801	137	0.132	32	0.283	129	0.646	217
1966	1.304	224	0.295	71	0.362	165	1.018	342
1969	0.973	167	0.266	64	0.496	214	0.986	331
1972	1.283	220	0.398	96	0.971	443	1.177	395
1975	1.559	267	1.896[a]	459	0.898	410	1.928	647
1978	4.017	689	1.829	443	1.688	771	3.561	1195
1981	6.170	1058	2.405	582	2.066	943	3.060	1027
1984	7.941	1362	2.248	544	2.870	1131	4.955	1663
1987	12.035	2064	4.755	1151	9.030[b]	4123	9.298	3120
1990	14.474	2483	5.975	1447	7.787	3556	10.555	3542

Notes: a. Very small acreage.-b. Large-scale, publicly financed irrigation schemes had begun to bring water to certain areas of Apulia in that year, thus sharply raising land prices (see also Apulia cooperative case study in Appendix).

Source: Calculated from public *Cassa* reports.

In addition, the regional development agencies, and their successors, the regional governments, were delegated by the *Cassa* to purchase and transfer about 23,000 hectares to about 900 families and 18 cooperatives under the provisions and funding of Law 590.

Interestingly, despite the *Cassa*'s original orientation towards the depressed areas, the single most important area for its operations in terms of farmland (38%), and particularly of value (53%), has been the North, in particular the Region of Emilia-Romagna (whose Po Delta, to be sure, had also once been a depressed area), followed by Central Italy, where the bulk of the operations (mostly in favor of former *mezzadri*) occurred in relatively well-off Tuscany. Only 35% of the acreage, and 27% of the value, of sales took place in the South and the Islands.

The true measure of the grant element resulting from the concessionary interest rates is the financial loss to date represented by the historical difference between the concessionary and market interest rates (including the present value of the *Cassa*'s end-1991 loan portfolio). This adds up to a grand total of 5,666 billion 1991 lire, or about US$4.7 [38]

Taking a round figure of 21,500 for the total number of *Cassa* clients to date (including an estimated 2,250 direct-beneficiary member families directly or indirectly benefited by land sales to cooperatives), it would appear that the total macro grant element of the 42 years of subsidized *Cassa* loans to date has been on the order of 264 million 1991 lire per client family, equivalent to about US$220,000.

At the micro level, the capital grant element implicit in the *Cassa*'s concessionary interest rate appears to have fluctuated between about 60 and 70 percent over the years, depending on the margin between the *Cassa* rate and the prevailing market interest rate (Table 6-4).

---

[38]The following are the underlying calculations (in million 1991 lire unless otherwise indicated:

1. Cumulative present value of capital transfers received by *Cassa* based on prevailing
   interest rate on government paper...6,417,862
   Less present value of 1991 portfolio, at actual interest of 4% and a discount
   rate of 7%, assuming a 24-year average loan maturity...............................- 752,267
   Total grant element... 5,665,595

TABLE 6-4: Capital Grant Element of *Cassa* Loans For Four Hypothetical Cases

	Case 1	Case 2	Case 3	Case 4
Year	1958	1976	1979	1991
Hectares	4	9	10	13
Price per ha.(000 lire)	300	3,300	3,750	10,500
Loan principal(mill.lire)	1.2	30	37.5	136.5
Cassa interest rate (%) (I_1)	1	2	4	4
Market interest rate (%) (I_2)	8.8	15.4	15.6	14.0
Ann.amort.@I_1 (000 lire)	47	1,340	2,170	7,900
" " @I_2 " "	115	4,685	5,925	19,500
Grant element [$1-(I_1/I_2)$]	0.59	0.71	0.63	0.59

NOTE: The first three lines represent averages of the *Cassa* operations for that year. The market interest rates, for lack of a more specific historical series, represent the average annual yields of long-term paper (source: Bank of Italy), plus an estimated average markup of 2.4 percentage points for lending institution's margin. The grant element (multiplied by 100) expresses a proportion of the loan principal.

It will be noted that the grant element was highest at the time the *Cassa*'s interest rate was doubled to 2%. This serves to explain why it was again doubled after only three years, evidently in an attempt to bring the grant element back down to what appears to be a "desired" level of around 60%. This may be compared with the European Community's capital contribution for land purchases for "restructuring" of marginal farms under its Mediterranean Investment Projects (PIM), of 45% until 1991 and 35% since 1992, applications for which appear to be chronically oversubscribed. (Data from an official report of the agricultural department of the Emilia-Romagna Region.)

A substantial number of borrowers (about 7,100 families and 36 cooperatives, for a total of 60,000 hectares, between 1972 and 1990) chose to pay off their loans after ten years. The average age of the loans, taking into account those who paid off early - 18 years in 1991 - seems surprising in view of the preceding calculations.[39] In line with earlier discussions (see Chapter 3), this behavior can be attributed to traditional debt aversion and to a desire for greater freedom to dispose of the land in case of subdivision among heirs (if approved) or to sell to third parties (See Fig. 6-1).

Another hypothetical exercise, it was thought, might indicate the "interest-paying capacity" of average family-sized (10-20 hectares) reporting farms in two ecological areas, and in two different periods, in each of the three regions that are representative of the *Cassa*'s operations, as shown in Table

---

[39] Investing, for example, in 3-months government bonds - BOT - which until recently yielded an after-tax annual interest rate of about 10%.

6-5.[40] The loan principal was assumed to be the current asset value of the land and fixed investments, as if the farm had just been purchased with a 100% Cassa loan. The maximum interest charge that the average farm family could bear in each circumstance was estimated by trial and error, applying various rates to the loan principal until a match was found for the "maximum amortization capacity". The ranges indicated for the latter in the last two columns are simply the differences between gross and net farm income, respectively, and the "poverty line". (It is important to bear in mind that the farm data represent averages of reporting farms, i.e., aside from not pretending to show changes over time on the same farm, they do not necessarily represent even the same farm sample over the ten-year period).

The real test, of course, is the long-term amortization capacity in relation to net farm income, rather than the short-term cash-flow capacity related to gross farm income, which includes the book value of depreciation charges. In the majority of cases, regardless of the Region, the values seem to fall into a range between 4 and 8 percent, with the minima and maxima related, respectively to net and gross income. Considering net farm income only, interest-paying capacity varies from one-half percent to 9 1/2 percent; only in the latter case -- representing efficiently managed farms in an industrialized region -- does it even approach (within about three percentage points) the prevailing market rate for mortgage interest. The Southern exception is assumed to represent irrigated vineyards and horticulture in the Apulian plains.

Figure 4.1 compares the trends of the official cost-of-living and farm price indexes with the index of average land prices paid by the *Cassa*, from 1960 to 1991. In general, land prices have moved up more rapidly than farm prices for most of the postwar years, the implication being that resource-poor peasants would have tended to find it difficult to pay market prices and/or interest rates for farmland (and also Fig. 6-2).

In the following chapter we will attempt to show, for a small number of real cases, *inter alia*, the implications of the nominal interest rate, *versus* a hypothetical, positive rate, for the per capita net farm income of farm-dependent members of the households.

---

[40] The farm accounting data were obtained from INEA's regional averages for family sized farms. These data are part of the EC's Farm Accounting Data Network (FADN), as reported by the appropriate regional volumes of INEA, *Strutture e redditi delle aziende agricole* for the years concerned. These are averages for representative numbers of farms in each region, i.e., they are statistical constructs rather than real situations. The poverty line reflects the official data for a family of three, respectively, in North, Central and South Italy. (The statistical "sensitivity" of the maximum interest rate capacity to increasing or decreasing the household members by one-half unit appears to be one-half percentage point). The 25% difference between asking price for the land and the Cassa-approved loans (see Chapter 4) should be kept in mind. Finally, the farm income figures tend to be understated in terms of the poverty line because (a) the family budgets on which the latter is based include the value of housing, whereas farm income does not, (b) farm income reflects only partially, or not at all, the value of home-consumed farm products, and (c) interest on indebtedness -- which may include loans for land purchase -- is subtracted before arriving at net farm income.

TABLE 6-5: Hypothetical Interest-Paying Capacity of Average Farms
in Three Regions, 1980-82 and 1989-91

Region Area & Year	Area Farmed (HA)	Labor Units (m/yrs)	Land Value	Net Farm Income	Poverty Line +10%	Maximum Amoritz. Capac.	Maximum Inter. Rate %
			------------------- Million lire -----------------------				
Emilia-Romagna							
Plains							
1980	13.9	2.7	366	22.7	14.3	8.4	0
1990	14.0	2.2	446	48.6	20.6	28.0	4.5
Hills							
1980	14.1	2.5	125	15.0	14.3	1.0	0
1990	14.4	2.3	352	60.4	20.9	39.5	10
Tuscany							
Plains							
1980	14.3	2.0	137	15.7	10.4	5.3	1
1990	13.9	1.5	205	24.0	13.8	10.2	3
Hills							
1980	14.0	2.4	90	11.2	11.6	-0.4	0
1990	13.7	1.9	181	25.0	16.5	8.5	2
Apulia							
Plains							
1980	13.7	2.1	135	34.1	7.7	26.4	20
1990	13.9	1.5	242	25.0	10.3	14.7	4
Hills							
1980	14.4	2.0	109	11.2	7.7	3.5	1
1990	15.2	1.3	221	19.0	9.4	9.6	2

Sources: Farm Data from Italian section of EC-FADN (average values for sample farms in size class 10-20 hectares). Poverty line calculated from ISTAT consumption survey for families of three members, unless otherwise stated.

Chapter 7

The case studies: A summary

Data and impressions collected from four cases of individual *Cassa* clients visited during one week of field travel are highly positive with respect to the socio-economic impact of the lending program at the individual level. Aside from the more generalizable, objective reasons discussed elsewhere, this is perhaps because all of the case study farms, selected by *Cassa* management, were operated by obviously entrepreneurial types, the "oldest" of whom got their loans in the early 1970s. Three were former *mezzadri* (i.e, long-term share tenants) on the farms subsequently purchased through the *Cassa*, all located in a fertile, industrialized part of northeastern Italy; the fourth was an equally entrepreneurial operator who had previously been a *mezzadro* in his home region of Marche and had already previously purchased some land in Tuscany without help from the *Cassa*. None of the farms visited were operated by former sharecroppers or landless farm laborers (except the case of the Apulia "cooperative" discussed below). However, a large, government-owned estate, studied in part in greater depth for another research project, was allotted in part to former laborers and employees.

All were characterized by a strong, traditional family cohesion, and the younger generation showed no signs of the desire of previous decades of farm youths to escape from the land. Indeed, one of the younger, managing males had studied vocational agriculture. With the exception of one person on one farm, only family labor is permanently employed, and casual labor is the exception even among the fruit growers. Investment in labor-extending farm machinery and equipment is high. The intensive fruit orchards (pears, peaches, apricots, apples, grapes and kiwi) of Romagna, which typically measured between 10 and 20 hectares, were using automatic drip irrigation equipment (mobile, in one case), and one was about to rent a grape picking machine.

* * *

The Tuscan "case" (township of Pienza in eastern Siena province) should be considered quite atypical although it is instructive in many ways. The farm enterprise actually consists today of two medium-sized farms belonging to two brothers who had migrated from the Marche in 1955, together with two more brothers and a nest-egg of 50 million lire[41]. They had bought 25 hectares in 1955 for 133 million lire (5.3 million lire per hectare), which they paid off in three years. In 1961 they bought another 18 hectares, of land of lesser quality, for 21 million lire (only 1.2 million lire per hectare). In 1976, they began to rent an additional 97 hectares, which they then bought, in 1984, for 592 million lire, or 6.1 million lire per hectare (389 million and 4 million at the current price), plus 3.4% for incidental costs, with a *Cassa* loan. Meantime, in 1977, they had already acquired a further 31 hectares for 337 million lire (nearly 11 million per hectare) with a 3%, 30-year loan from a regional agency.

At the time of the visit, the two brothers asserted that they were farming a total of 364 hectares consisting mostly of hillsides devoted to coarse grains and grassland. Originally, they farmed jointly, but later they decided to divide the farm in two for management purposes, although they continue in joint ownership, collaborate closely and share some of the financial aspects, such as the loan repayments. Both

---

[41]All lire figures in this chapter (unless otherwise indicated as in the case of the amount of the *Cassa* loan) are expressed in 1991 prices. They may be converted to approximate 1991 US dollars by dividing by 1,200.

are specializing in confined raising of Charolais bull calves purchased in France for sale throughout the year as yearlings for slaughter; there were about 700 head on both farms at the time of the visit. This system had largely replaced the traditional, diversified Tuscan land use pattern of yesteryear that would have required far more than the total of 2 1/2 man/years of total permanent labor input at this time, including one hired man (an East European; there appears not to be any indigenous, permanent-hire farm labor available in the areas of Tuscany and Romagna that were visited).

Considerable investments had obviously been made in the improvement of the homesteads, one of which was a former *mezzadro* residence, the other was the residence of the former owners of the estate from which the farms had been carved. Personal vehicles were of fairly recent vintage and furnishings resembled those of well-off urban middle class.

One of the two brothers admitted under questioning that, given land/produce price relations in the early 1980s, he would have been willing to pay 8-9 percent interest on his mortgage loan. Indeed, partly under pressure of demand from urban investors and domestic and foreign buyers of residential estates, average farm land prices in Tuscany paid by the *Cassa* in 1990 were three times their 1984 level.

*   *   *

Three farms were visited in the province of Ravenna in eastern Emilia-Romagna region. All were operated by families that had been *mezzadri* on the same *podere* that they subsequently purchased from the former landlord, with loans from the *Cassa*. All three are fruit growers in areas that had specialized in such crops for decades and all had intensified land use, and enhanced farm income, tangibly since they became owners. Farm accounts for the past three years were made available by the technical assistance and accounting service[42] to which they all subscribe. A farm management analysis was available for only one of the three farms.

The per capita income data for these cases are compared, respectively, to the following <u>official annual average per capita</u> parameters for 1991 for Northeast Italy (from ISTAT 1992):

		Million lire	(1000US$)
o	Income, all households	14.6	(12.2)
o	Expenditure " "	13.7	(11.4)
o	Income, agricultural households	12.4	(10.3)
o	Expenditure " "	11.1	(9.3)

---

[42] The Technical Assistance Center of CISL-Coltivatori, a small-farmer-oriented offshoot of a labor union, located in Lugo (Ravenna).

They are also compared to the national poverty line, by applying to 1990 household expenditure data the coefficients developed by the survey commission on poverty and marginality in its second report in 1988[43].

$*$ $*$ $*$

GG.'s farm in the township of Solarolo comprises nearly 11 hectares of bottom land purchased through the *Cassa* in 1976 for 376 million lire of 1991, or about 34 million per hectare (75.6 million and 7 million, respectively, in 1976)[44] in addition to 15 hectares of rented hillside. (For health reasons, the owner has had to turn over active management to his son -- a surveyor -- and they wish to purchase, and plant to fruit trees, the hillside lot in order to be able to occupy the son-in-law, as well, full-time in the family enterprise). In 1991, they paid 8.5 million lire for occasional outside labor, equivalent to barely 11 percent of net farm income.

The annual amortization to the *Cassa* is 3,583,000 lire, equivalent to about five percent of the 1991 net farm income of 72.4 million and, in the operator's words, little more than the rent he was paying as a *mezzadro*. The interest charge of 1,083,000 lire was only 1.5% of net farm income. (Average net farm income for the three years, in terms of current lire, was seven percent greater than the face value of the *Cassa* loan.)[45] Net 1991 farm income per man/year of operator's and family labor (est: 2 man/years) was 36.3 million lire; for the four household members depending on farm income, it was more than three times the poverty line and 1½ times the average for agricultural households in NE Italy. (Depreciation allowances totaled 20.6 million.)[46]

---

[43] *Commissione d'indagine sulla povertà e l'emarginazione,* <u>Secondo rapporto sulla povertà in Italia,</u> Milano, F. Angeli 1992, pp. 28-29. It applies the international standard to Italy's annual household budget survey. Annual per capita consumption in 1991 of 13,680,000 lire per year was accepted as the poverty line for a family of two in Northeast Italy, as shown below. (Note that this figure includes expenditure for housing (15%), which does not figure in the farm household's budget.)

Family size	Equivalence coefficient	Poverty line, 1991 (1,000 lire per year)
1	60	8,316
2	100	13,860
3	133	18,433
4	163	21,679
5	190	25,270
6	216	28,728
7 or more	240	31,920

[44]In this and the following case analyses, it should be borne in mind that, as pointed out previously, the *Cassa*-approved price, and hence the loan value, have tended to be about 25% below the asking price.

[45]The enterprise was at the time also amortizing three highly subsidized emergency loans for "atmospheric calamity" (hail damage) and one for machinery purchase, for a total annual cash outflow of 18 million lire, which, however, included less than 2 million in interest.

[46]It should be borne in mind that in all cases analyzed here, the charging of a depreciation allowance, as required by modern farm accounting, does understate the actual cash availabilities.

Net operating income in 1990 was even higher, at 80.9 million[47], equivalent to 40.5 million per man/year of labor input and about 20 million for each "consumption unit". By contrast, as a result of heavy hail damage to the fruit crops, net operating income in 1989 was only 31 million lire. This still yielded a respectable 15.5 million per man/year of labor input but only 83% of the regional average per capita expenditure.

The average net operating income for the three years (61.4 million) for the farm-dependent household was 24% above the average per capita income for the region (from which an average of 15% was spent for housing), and 2½ times the poverty level.

If GG had been paying interest on his *Cassa* loan at an assumed market rate of 12%, his 1991 interest payment would have amounted to 8 million lire (implying a subsidy of 6.9 million lire). His total amortization payment would have been 9.5 million, equivalent to 13% of net farm income, and implying a grant element of 62% with respect to his actual annual amortization payment of 3.6 million lire. The resulting net farm income of 62.9 million would have still provided 2½ times the poverty line level of expenditures for a family of four, and one-fourth more than the per capita average for NE Italy farm households.

In conclusion, this borrower, mid-way through his *Cassa* amortization, could have well afforded to pay the market interest rate at that time, even in the face of his 1986 calamity and his other financial commitments.

*   *   *

BG.'s 11-hectare farm in the township of Brisighella (near Faenza) was bought with a Cassa loan in 1984 for 236 million lire of 1991, or 21.5 million per hectare (155 million and 14 million lire, respectively, at the current price); the farm consists half of well-drained, irrigated bottom land and half of hillside.[48] The family had farmed the same plot as *mezzadri* for 20 years. The household consists of six persons but only the 54-year-old owner-operator works the farm full-time. The 17-year old son was still in school but was hoping to help run the farm soon after graduation. However, it seemed that he would have to follow his brother-in-law's example for a while and seek work off the home farm, helping out only in his spare time. BG asserted that he is interested in buying six adjacent hectares, but more because of conflict with its present occupant than for economic reasons.[49] The investments made since the purchase consist mostly of having converted to fruit tree plantings three hectares formerly devoted to field crops and olive trees. (More than half the farm continues to be planted to wine grapes which are pressed on the farm.) The farm accounts do not show any indebtedness other than to the *Cassa*.

Annual amortization of the Cassa loan amounts to 7.8 million lire, of which 3.3 million represented interest in 1991. The total was equivalent to a sizeable 44 percent of net operating income. The interest charge alone was 18% of net income. Net operating income, after deduction of all depreciation charges,

---

[47]The figures for 1990 and 1989 are stated at current prices; the CPI increased by 6% in each of the two years.

[48]This is reflected in a price per hectare (in 1991 lire) only two-thirds of that paid by GG in 1976, as well as in the farm's productivity per hectare.

[49]An indication of the sharp appreciation in land values since the farm was purchased is the asking price for this plot: 50 million lire per hectare, or 3 1/2 times the price paid only eight years earlier, while the CoL index rose only about 60%.

was only 17.8 million lire, equivalent to about 12 million per man/year of family labor input (estimated at 1.5 man/years), and about 6 million for each of the three household members assumed to depend on the farm income. This was only half the average per capita regional farm family income, and it nearly coincides with the poverty line. (Depreciation charges amounted to 20.1 million lire for the year.)

Farm income can, of course, be highly variable from year to year. In this case, net operating income in 1990, at 31.5 million lire, was about three-quarters higher than in 1991, while in 1989 it was one-third less, at 12.6 million, i.e., only 8.4 million per man/year and 4.2 million per consumer. Average net farm income for the three years was 20.6 million lire. Even at the subsidized interest, amortization to the *Cassa* was equivalent to a respectable 38% of that average, and interest to 16%.

If interest had been charged at the market rate, the year 5-7 average interest payment would have amounted to about 18 million (implying a subsidy of 15 million), and total annual loan amortization would have risen to over 19 million, resulting in an average net operating income of barely 1½ million lire.[50] (The implied grant element was 59%.)

In this case, the operator would have clearly defaulted on his loan in the early years at any interest rate substantially in excess of the subsidized rate of 4%.

*　*　*

The L. brothers in Bagnocavallo purchased their fully irrigated *mezzadria* holding of 18 hectares for 440 million lire of 1991, or 24.4 million per hectare (289 million and 16 million, respectively, at the current price), also in 1984. With 11 hectares already in fruit trees and grapes, calamity hit in 1986 in the form of a devastating hailstorm that wiped out most of the crop and even necessitated replacement of part of the fruit tree plantation. They survived mainly by virtue of a government disaster loan of more than 40 million lire bearing only one percent interest, and by the sale of a 30-head herd of dual-purpose cattle that they had already owned as *mezzadri*. (The cattle sale actually induced the L.s to convert all of their six hectares of field crops to fruit plantations with the help of the special loan program.)

This household is composed of 10 members, including the two younger brothers who, with their respective spouses, own the farm (the other two, both single, were already officially pensioned at the time of purchase; they could thus not be officially classified as active farmers and were therefore not entitled to co-sign for the *Cassa* loan). Three women of the younger generation work off the farm (which must have also helped overcome the hail losses). The only male heir is today the most active of the members working on the farm, having recently graduated from a vocational agricultural school. While seven members of the household share in the farm work, the total equivalent of family labor is given as five man/years in the farm account analysis for 1991; no outside labor is employed even at harvest time. For purposes of calculating per capita farm income, seven members of the household are assumed to depend totally from it, although the others can be assumed to contribute part of their off-farm earnings and

---

[50]The farm was also beginning to pay off a 10-year, 5.7% interest home improvement loan of 48 million lire to the official, regional farm credit agency, at the rate of nearly 5 million per year. What probably happened was that the household members working off the farm would have chipped in to make ends meet. (A complete socio-economic analysis of this kind, of course, should comprise total household income and expenditures, in addition to the farm accounts.)

pensions to the household budget (see above). The families evidence a cultural level that belies their "*contadino*" background.[51]

Net farm income averaged 70.9 million lire for the three years, 1989-1991. It was highest -- at 81.2 million -- in 1991. Amortization of the *Cassa* loan amounted to 15.7 million each year, of which 6.2 million was interest. This is equivalent to 22% and 9%, respectively, of the average net farm income. (The average reserve for depreciation was about 24 million.) Amortization of the disaster loans and other medium-term liabilities absorbed an additional 6 million, raising total cash outflow on account of loan amortizations to the equivalent of 28% of net farm income.

The average net farm income per man/year of family labor input for the three years was 14 million lire. Net farm income per farm-dependent household member was more than 10 million, or 80 percent of the average for the region but 2.2 times the poverty level.

If the interest rate on the *Cassa* loan had been at the market rate, interest payments to the *Cassa* alone would have amounted to 33.8 million (a subsidy of 27.6 million), and the annual amortization would have been 35.9 million. This would have reduced the average net farm income for the three years to 35 million, a mere 7 million per man/year, and only 14 percent above the poverty line for a family of seven. (The implied grant element was 56%.)

Comparative analysis of the farm accounts (available only for this farm) shows that, despite the enormous loss from hail in 1986, a few years later it was nearly 50% above the average farm in its group in the region in terms of returns to operator's and family labor (13.6 million lire per man/year in 1991) as well as in terms of net farm income per man/year and per standard hectare (70% and 31%, respectively, above the average). Good management was also evident in the total cost per hectare which, at 3.9 million lire, was only 78% of the average farm's. This despite the fact that the enterprise paid 532,000 lire in interest charges per hectare, or six times as much as the average farm. (With the *Cassa* loan at the market rate, interest cost per hectare would have exceeded 1.5 million.)

* * *

The two cooperatives that were visited illustrate to some extent the socio-cultural and economic gap that still exists between North and South and that is masterfully documented and analyzed by Putnam. The gap is reflected largely in the fundamental -- and not at all atypical -- difference in the nature of the two organizations. The first is located in the cradle of cooperative joint farming in Italy (dating back to the end of the last century and based on solidarity values of both Christian and Socialist origin) among landless laborers who had been employed in the reclamation of the land that they later farmed. In the more specific case of the Region of Emilia-Romagna, moreover, one can speak of a deep-rooted cooperative culture which has proved largely compatible with modern forms of organization.[52]

The second case, by contrast, situated in an area devoid of a cooperative culture, represents a cooperative created for formal reasons, a kind of Hobson's choice as an alternative to a fully privatized

---

[51]The dwellings are rather expensively, and tastefully, furnished, and the two elder brothers have impressive art collections. The families are related by marriage to the director of a well-known French music conservatory.

[52]See also PUTNAM, p. 144.

solution for the estate which would have sharply reduced the work force. This purely nominal cooperative owes its existence solely to the overwhelming desire of the former laborers to preserve the privileged status they enjoyed as employees of a public sector conglomerate. (For a summary of these two cases, see Appendix).

*   *   *

A fairly spectacular case illustrating the *Cassa*'s usefulness is offered by the recent, partial subdivision of a large, State-owned farming estate on the outskirts of Rome, known as "Maccarese".[53] The originally privately developed estate of 3,600 hectares was rescued from bankruptcy in 1933, along with many other large Italian firms during the Great Depression, by IRI, the giant public holding company created for this purpose. IRI had been trying to run it ever since like a plantation with a labor force composed largely of former construction workers brought from the Veneto region for the original reclamation project, and their descendants. After decades of wrangling with the well-entrenched and politically organized workers -- who numbered 872, or one for each 3.2 hectares, in 1975 -- the enterprise has never been able to show a profit.

Several attempts to privatize the property over the years failed, largely because potential buyers could not come to terms with the organizations representing the partly redundant labor force. In 1980, IRI finally declared the enterprise to be bankrupt, and a receiver was appointed. Meanwhile, about 100 of the 120 families who had agreed in the 1950s to farm 800 hectares as share-tenants (*mezzadri*), accepted an option to buy about four hectares apiece in the 1970s on easy terms, for about 13 million lire of 1991 per hectare. (These sandy farms, well-adapted for growing irrigated produce for the Rome market, have reportedly done very well.) Following a failed attempt to convert some of the workers (without individual farm management experience) into sharecroppers, and one to sell the enterprise to a cooperative of all the workers (opposed by all but 80), the *Cassa* was called in, for the first time, in 1982, to help implement a plan for a rationally designed sell-off of a substantial part of the estate to those workers who were willing and able to take on independent farm management and the accompanying financial and other legal obligations attached to *Cassa* loans. The lots were sized in accordance with soil quality (ranging from sandy to clayey and from 3 to 7.5 hectares) and sold for 40 million current lire (77.5 million at 1991 prices), with an option to buy a dwelling for an additional 11-36 million current lire (21-70 million lire of 1991). The offer was taken up by barely 50 of the remaining 315 employees. Despite some initial difficulties due to management inexperience, 40 of the original holdings were found to have succeeded in providing full-time work and an acceptable income level for the families after ten years, particularly those located on sandy soil.[54]

---

52 The general information is summarized from Margaret Loseby and Pierdomenico Ceccaroni, <u>The transformation of a large state-owned farm in Italy</u>, paper presented to the FAO-ECE Workshop on Specific problems of the transformation of collective farms into viable market-oriented units, Godollo, Hungary, 22-26 June 1992.

[54] The balance of the property is included in the government's current high priority for calls for privatization tenders. At this writing the Italian government has various bids from domestic and foreign investors, some of whom appear to be interested mainly in the enormous potential for non-agricultural use of part of the property.

49

Chapter 8

Conclusions

Conclusions regarding the future of the *Cassa* and its operations in Italy are expected to be dealt with in the Italian version of the report and in the context of the outlook for Italy's and the EEC's agricultural and rural policies. This chapter attempts to formulate some conclusions regarding aspects of the Italian government's measures in support of the land market for family farm formation and consolidation, and in particular the *Cassa* experience, that may be of relevance for certain developing countries. We believe that they are particularly relevant for most of the countries of Latin America (including those which have, over the decades, steadfastly denied the need for "land reform"), and to many of the countries of the former Soviet Union and of Central and Eastern Europe within their own temporal and institutional contexts, as they struggle with the conundrum of how to re-privatize once-reformed, collectivized and/or statized land tenure systems. **Obvious lessons of an institutional, procedural and economic nature** can be drawn from the singular Italian experience. Above all, we have here a rare example of specific institutional and financial measures for the **sustained implementation of an explicit government strategy in favor of the family farm via the land market.** But these measures were effective in the early years in large part because they were accompanied by a drastic, quick, coercive agrarian reform which, among other beneficial effects, induced many owners to put medium and large-sized estates on the market both prior to the enactment of the reform legislation and subsequent to its implementation.

*   *   *

o   On the basis of the various references in previous chapters to Italy's political and socio-economic context since World War II, there can be little doubt that **Italy's policies and measures in support of the family farm were fully justified.** Such a judgment is, of course, reinforced by the entire European Community's explicit choice of the family farm as <u>the</u> land tenure system to be backed by a multitude of incentives and expensive subsidies.[55] In the case of Italy, even in pre-EC years, the most manifest expressions of its resolve to promote and assist the expansion of the *proprietà coltivatrice* were, in logical, if not quite chronological, order, (a) the creation of more than 100,000 new peasant owners on 600,000 hectares of land expropriated from 3,300 *latifundia* in the regions of greatest socio-economic contrasts, (b) the constant (if indirect and perhaps unintentional) pressure on rentier landowners to divest themselves of their holdings through repeated legislative "reforms" of cash and share-tenancy contracts, and finally, (c) the heavily subsidized long-term farm mortgage loans targeted explicitly and exclusively on the small farmers and farm laborers, including the program managed by the *Cassa*. The flexibility of the mortgage loan programs in terms of the size of newly constituted or enlarged family farms -- as reflected in the gradual increase of their average size over the years -- has allowed the programs to survive the drastic changes that have taken place in family farmers' income expectations, off-farm employment opportunities (including part-time farming), relative land prices and farm capital requirements.

*   *   *

---

[55]"The very balance of our societies is threatened by the decline in agricultural employment, the growing split between town and country ... the departure of the young and, finally, the damage done to the environment." Thus (apparently even in the face of these policies) writes Jacques Delors, President of the CEC, in *Le Figaro* as these conclusions were being drafted, as quoted by The Economist of August 14, 1993.

50

o   The *Cassa*'s concessionary interest rate, its readiness to finance in principle 100% of the nominal price (and in fact an average of about 75% of the real price) and its objective vetting of that price, as well as the transaction tax and fee breaks, were presumably designed to allow deserving aspirants to family farm ownership to compete in the land market who might otherwise not have been able to do so. Indeed, these 45-years-old **measures go to the core of the conclusions reached by Binswanger, Feder and others in recent years for developing countries, i.e., that** resource-less **peasants cannot be expected to acquire farmland at market prices and market interest rates through any kind of "reform" mechanism** because, even in the case of long-term loans, the amortization payments inevitably tend to reduce family income below the subsistence level.[56]

*   *   *

o   **The principle of the interest subsidy -- or other forms of grant, such as a subsidy for the land price -- for this purpose**[57] **can thus be considered valid,** given the socio-political need for stability in the rural areas. Nevertheless, there remains the question of **whether the policy objectives could have been attained at a lower cost.** In other words, how much of the enormous grant element, within or without the *Cassa*, has been justified by the policy objectives, on the one hand, and, in retrospect, by the macro-economic and land market realities? On the basis of the trend of peasant purchasing power in terms of land, and the scanty evidence provided by our few, and unrepresentative, case studies (Chapter 7) and by the hypothetical cases (Chapter 6), one could conclude that, **in retrospect, the generalized subsidy has been excessive.** This is felt to be true at least at the current, long-time level of 8-10 percentage points below the market rate (i.e., borrowers have been paying in effect only about one-third of that rate) when dealing with entrepreneurial family farmers, especially when they seem to be able to finance up to 25% of the purchase price from savings, and in a period of buoyant farm and land prices.[58]

---

[56]BINSWANGER *et al* (1993) summarize the dilemma as follows (p.52):

"Consider an idealized case of competitive and undistorted land, labor and credit market. The value of land for agricultural use would be equal to the present value of agricultural profits capitalized at the opportunity costs of capital. If the poor have to use credit to buy land at its present value, the only income stream they have available for consumption is the imputed value of family labor. They must use the remaining profits to pay for the loan. If the poor can get the same wage in the labor market, they are not any better off as landowners than they would be as wage laborers. This example is, moreover, an ideal situation where the interest rate paid by the poor is equal to the interest rate that the most credit-worthy borrowers can obtain.

If, as discussed above, land ownership provides access to credit and helps in risk diffusion, the buyer has to compensate the seller for the utility derived from these services of land, which will be capitalized into the land value. Feder et al (1988) develop a model identifying the role of credit subsidies in land price determination. Only unmortgaged land provides these services. A buyer relying on credit cannot pay for the land out of agricultural profits alone but must have either prior savings, or must cut consumption below his (imputed) market income. Anything that drives the land price above the capitalized value of the agricultural income stream thus makes it impossible for the poor to buy land without reducing consumption below what they could earn in the labor market. Several other income streams associated with land ownership are usually capitalized into land values."

[57] *Pace* the World Bank's (still-debated) principle that other types of farm credit should pay market rates.

[58]An important factor to take into account in the Italian equation is the exceptional phenomenon of rising real farm wages that also vastly outran the trend of land prices through 1980, as outlined in Chapter 4. For our purposes, it represents a two-edged sword: (1) the levels of "subsistence" and "poverty" assumed for recent years are inflated; (2) so are the cash labor costs of the entrepreneurial family farmer. Neither would be appropriate for analysis in the countries to which these conclusions are addressed.

51

Given that *Cassa* borrowers pay the market price for the land, the case studies and hypothetical examples also indicate that a perhaps significant proportion could not have paid a higher interest rate without seriously depressing their level of living. Since demand for *Cassa*-type financing, as for any type of credit, is obviously price- (i.e, interest rate-) elastic, this means - *ceteris paribus* - simply that there would have been less competition for these loans from the more risk-averse among the potential clients and for marginally less productive land.

A final consideration related to the question of how much subsidy is required to attain the desired policy objectives, under conditions similar to those of Italy over the past 45 years, is the grant element implicit in the substantial rise of real land prices (i.e., compared to general price inflation) that took place during certain periods. The fairly large number of borrowers who paid their loans off early (after the trial period) may well reflect, *inter alia,* general awareness of the windfall profits to be made from resale of the property.

*   *   *

There is an additional public policy consideration that is not always given the attention it deserves. Public subsidization of structural tenure changes through the market, i.e., at or near market land prices, is obviously limited by public financial availabilities. In the presence of many aspirants, as in postwar Italy through the 1950s, it is not only politically difficult (except under authoritarian regimes) but also of doubtful rationality to earmark public resources for a few in order to create islands of "modern agriculture" with incomes comparable to, or greater than, those in other sectors. The question then arises "what about those who remain on the outside"? Under conditions of lesser population pressure on the land and greater inter-sectorial integration, as in Italy beginning in the second half of the 1960s, different criteria may rule. In fact, the *Cassa*'s evolution over time reflects these changes in the general socio-economic context.

*   *   *

o   Each country wishing to implement a similar strategy will need to decide for itself, based on local conditions and considerations, not only what kind of transactional and financial support (with a minimum of bureaucratic interference) would best serve the establishment of viable family farms,[59] but, above all, the magnitude of the grant element required. Ideally, such a decision should be based on careful study of <u>real</u> farm and family-level data, of the existing land market and of the price of suitable farmland in relation to the land's economic potential at prevailing farm product prices. What is certain is that **few, if any, countries that may be concerned will be able to afford the generalized level of subsidy that**

---

[59]One could try to analyze the advantages and disadvantages, in terms of cost and of expanded coverage, if the *Cassa* had borrowed its investment funds in the capital market, with the government picking up only the interest rate subsidy and administrative costs, or (b) had merely performed its non-financial functions plus one of guaranteeing, on behalf of the government, subsidized mortgage loans for 100% of the approved price made by the banking system, rather than operating with capital transfers from the central government budget. However, in view of the weakness of the banking system in most of the countries where such a scheme may be initiated today, and of the proven lack of private bank interest in Latin America, for instance, in financing peasant farmers and in long-term lending in general, such questions are largely academic. Also, the issuance of special debt instruments for financing the scheme could be considered if a market for mortgages exists or can be created. It may be interesting to trace the use the *Cassa* makes of the new funding authority introduced in 1991 (see Chapter 5).

52

has been ascertained in Italy, particularly if they cannot guarantee an efficient system of payment collections and of property foreclosure in case of willful loan delinquency.

No magic formula can be proposed for this purpose. At most, one can suggest the obvious: that **the subsidy be limited in principle to keeping the amortization of the land price from (a) depressing family income below a reasonable subsistence level, and (b) leaving a margin for additional indebtedness for capital investment (which, in turn, may require some subsidy element).** For equity reasons, as well for reasons of economic efficiency and for administrative convenience, it can be argued that the subsidy ought to be uniform for all applicants. Whether its level ought to be based on worst-case or average farm management assumptions within the existing input/output price structure, is a matter for policy makers to decide. They will have to weigh -- within the limits of public financial resources -- whether the emphasis should be on allowing the least able to survive (**equity considerations**), while providing a perhaps well-deserved "rent" to the better managers (in the expectation that a large share of this rent will be reinvested), or whether **efficiency criteria** should prevail so that only the average or better farm managers would apply given the level of amortization payments for which they would have to obligate themselves.[60]

On the other hand, it can be argued that different types of borrowers[61] need different kinds and levels of subsidies. In the case of Italy, the land and cash-poor southern *bracciante* of the immediate postwar era who aspired to become owner-operator of a parcel of land doubtless required a greater degree of subsidy than the experienced *mezzadro* of central and north-east Italy with tangible savings in cash and kind who wished to convert himself from half-share tenant to independent operator; yet, both had to compete under identical grant levels. This would help explain the reportedly small proportion of *braccianti* who participated in the *Cassa*'s program even in the early years: as in the case of other types of uniformly subsidized, and untargeted or imperfectly targeted, farm credit, the better-off farmers almost inevitably garner the lion's share of the available funds. Similar considerations could be postulated in an inter-temporal sense, for instance, as between the typical *mezzadro* of the 1950s and his modern, highly capitalized and EC-price and income-supported counterpart (see Chapter 7).

*   *   *

Finally, a word about the restrictions imposed by Italian law on the alienation and subdivision of the properties acquired with government subsidy, as well as about the qualifications required for applying for the subsidies (the restrictions and qualification are similar to those to which land reform beneficiaries were subject). They were obviously designed essentially to discourage speculative demand for these loans in a rising land market (see above), and they appear to have been effective.

The main rationale for the <u>30-year</u> restriction on subdivision (except in case of succession), on the other hand, may well have been an attempt to circumvent the lack of a basic farm inheritance statute (similar to the German *Erbhofgesetz*) designed to prevent socially undesirable farm fragmentation.

---

[60]Whatever the level of subsidy, such a scheme presupposes an efficient and strict collection scheme which does not countenance loan delinquency.

[61]The use of the term "borrower" (rather than, for example, "beneficiary" or "buyer") is intentional. In the first place, we are firmly convinced that, for both social and economic reasons, land should not be distributed free. Obviously, we are equally convinced that suitable credit facilities must be created for this purpose.

53

Alternatives might be considered that would be specifically targeted on the prevention of speculative resale, such as land use restrictions in the title to any parcel alienated from the original property that would limit its use to agricultural or environmental purposes for even longer periods.

The qualification of *bona fide* family farmer or farm laborer for entitlement to acquire a subsidized farm property is essential lest a new class of subsidized, urban-based land *rentiers* be created.

APPENDIX

Two cases of "cooperatives"

The bottom line in the case of the huge cooperative (traditionally affiliated with the Communist party but today devoid of dogma), headquartered in Argenta (Ferrara province) in the lower Romagna, is that it is a dynamic, multi-purpose enterprise with a management hardly distinguishable from that of a profit-making firm (except for its distinct social orientation and environmental awareness). The successive *Cassa* loans, amounting to 11.4 billion lire for 938 hectares purchased over nine years beginning in 1984, that it has contracted for acquiring assorted farmsteads (including those of a group of former land reform beneficiaries who couldn't make it on their own)[62] have helped it prosper and expand, and its diversified income sources have prevented the loan amortizations from becoming a cash-flow problem.[63] Interest on all mortgage loans for land purchases (including debts to other lending institutions at interest rates ranging from 1 to 4 percent) in 1991 amounted to 389 million lire, barely 11 percent of total fixed costs and 1.5 percent of total turnover.

The Apulia "cooperative" in the township of Margherita di Savoia, a beach resort, on the other hand, is essentially a group of former hired hands of a large, intensively field- and vegetable-cropped estate whose owners -- a State-owned food and agriculture conglomerate originally based on electric power[64] -- decided to divest themselves of their agricultural holdings in 1987. (In order to avoid problems with the existing -- and probably redundant -- labor force, the holding was offered to them as a "cooperative" with the intermediation of the *Cassa*.

In 1992 the cooperative had 29 members, of whom 27 were farm workers and two were technicians. When the *Cassa* approved the loan, it stipulated that the manager for the previous 15 years be retained. Nevertheless, largely owing to labor redundancy (and the lack of local employment alternatives)[65], the cooperative is not producing a surplus for investment (not to speak of distribution) after paying the enormous labor bill and the annual amortization of the *Cassa* loans, in addition to the other fixed and operating expenses.

For 1991, the cooperative's accounts show a surplus of barely three million lire, against an accumulated net loss of 346,000 lire from previous years. (The net loss in 1989 was reportedly 180

---

[62]Most of the reform beneficiaries in this area were former occasional rural laborers without any experience in managing a farm. For political reasons, the reform agency had refused, in the 1950s, to allot the land in joint ownership to cooperatives of these laborers. However, in 1975 they succeeded in transferring, with the help of the *Cassa*, ownership of some plots to the precursor of the cooperative, for the first ten years in the names of individual members. The land is now irrigated.

[63]The cooperative had gross assets valued at 64 billion lire at the end of 1991, and a turnover of 25.6 billion for the year, which included 4.4 billion in wages paid to the some 600 "active" members, who constitute the entire labor force. Aside from sales of farm produce (some of it processed into alfalfa meal), earnings come from a such recently formed activities as construction contracting (2 billion lire grossed in 1991) and agro-tourism; the coop is also branching out into land development in east-central Italy and in Portugal.

[64]Parts of it are currently in process of being privatized.

[65]Note that the *Cassa* buys only land and fixed assets. In this case, the labor force would become even more redundant if operations became more mechanized, as sound farm management would require in view of today's price/cost squeeze.

million, for an average of 5 million per member, after deducting the previous year's surplus.) Against gross assets of 6.3 billion lire, the balance owed to the *Cassa* amounted to 5.1 billion. This included 266 million for overdue payments; no amortization payments were made in 1991, and only 193 million are reported for 1990. On the other hand, against a total turnover of 2.5 billion lire, the total wage and salaries bill in 1991 was 905 million, of which 656 million was paid to members (including social security contributions), 108 million was for administrative salaries and contributions; an additional 32 million was spent for technical consultancies. The cost of labor for an average of 30 workers, day in-day out, is calculated to average 17,000 lire (about US$14) per hour, a charge that this type of enterprise cannot bear, especially at a time of falling produce prices for vegetables owing to foreign competition which has already caused some private farms in the area to fail.

Although the members had agreed, in view of the 1989 debacle, to work free-of-charge for 200 hours a year each, this turned out to be insufficient; moreover, half the members were against the "sacrifice". According to the president, most now agree to stop the present wage-based system, but the opposition is militant and there have been cases of evident sabotage.

The fundamental problem is that the group is suffering from collective dual personality: on the one hand, they are a legally registered "cooperative", and purchased the estate as one, while, on the other hand, they also have a registered, collective labor contract with their own cooperative and thus expect to continue not only to enjoy the benefits of that status but also to be fully employed throughout the year at the market rate, regardless of the labor requirements of the enterprise under competitive management.

REFERENCES

BARBERO, Giuseppe *"La dinamica delle strutture e la pianificazione territoriale"*, <u>Rivista di Economia Agraria</u>, 1965 No.2-3

---   *Tendenze nell'evoluzione delle strutture delle aziende agricole italiane*, INEA, Rome 1967.

BINSWANGER, Hans, Klaus Deininger and Gershon Feder, "Power, Distortions, Revolt and Reform in Agricultural Land Relations", in *Handboook of Development Economics*, Vol. III, Jere Berman and T.N. Srinivasan, eds., published in 1994.

CAMPUS, F.   *"La dinamica delle strutture e della occupazione del suolo nelle zone ad economia depressa"*, <u>Rivista di Economia Agraria, 1965 No. 2-3, .</u>

CASSA PER LA PROPRIETA CONTADINA, <u>I Terreni Acquistati dal 1948 al 31 Dicembre 1990; Le Cifre per Regioni</u>, Roma 1991.

---   <u>*Statuto*</u>, Roma 1983.

FANFANI, Roberto, <u>*Lo Sviluppo della Politica Agricola Comunitaria*</u>, *Roma, La Nuova Italia Scientifica*, 1990.

GIACOMINI, C., "*La politica di riordinamento fondiario in Italia e gli interventi per lo sviluppo della proprietà coltivatrice*", in <u>*I problemi finanziari dei giovani agricoltori*</u>, Milano, Dott. A. Giuffrè, 1982.

INEA -*Istituto Nazionale di Economia Agraria*, <u>*Annuario dell'Agricoltura Italiana*</u>, various years.

ISTAT - *Istituto centrale di statistica*, <u>*La distribuzione quantitativa del reddito in Italia nelle indagini sui bilanci di famiglia, anno 1985*</u>, *Collana d'Informazione* , n. 25, 1992.

LORENZONI, G., <u>*Inchiesta sulla proprietà coltivatrice formatasi nel dopoguerra, Relazione finale: L'ascesa del contadino italiano nel dopoguerra*</u>, Rome, INEA 1938.

MCENTIRE, Davis, "Land Policies in Italy", in McEntire and Agostini (Eds.), <u>Towards Modern Land Policies</u>, University of Padua, 1970 (?).

MINISTERO DELL'AGRICOLTURA E DELLE FORESTE, <u>*Relazione al Parlamento sul primo e secondo Piano Verde*</u>, Roma 1968.

PUTNAM, Robert D., <u>Making Democracy Work, Civic Traditions in Modern Italy</u>, Princeton University Press, 1993.

ROSATO, Paolo, "*Un'analisi del mercato fondiario veneto: i fattori che influiscono sul prezzo dei terreni agricoli*", <u>*Genio Rurale*</u>, No. 2, 1991.

ROSSI DORIA, Manlio, <u>*Analisi zonale dell'agricoltura italiana mediante disaggregazione dei dati regionali*</u>, mimeo, Portici 1965.

SHEARER, Eric, Susana Lastarria-Cornhiel and Dina Mesbah, <u>The Reform of Rural Land Markets in Latin America and the Caribbean: Research, Theory and Policy Implications</u>, Madison, Wisconsin, Land Tenure Center, University of Wisconsin, Paper No, 141, April 1991.

ADDITIONAL BIBLIOGRAPHY

ADORNATO, Francesco, *"Un Nuovo Modello di Cassa per la Formazione della Proprietà Contadina"*, *Nuovo Diritto Agrario*, XV, No. 2, 1987.

BARBERO, Giuseppe, *Riforma Agraria Italiana*, Milano, Feltrinelli, 1960.

---   <u>Land Reform in Italy</u>, Rome, FAO, 1961.

---   e Giuseppe Marotta, *"Mobilità e Mercato del Lavoro Agricolo dal Dopoguerra ad Oggi"*, *in Piero Bevilacqua (Ed.), Storia dell'Agricoltura Italiana in Età Contemporanea, II Uomini e Classi*, Venezia, Marsilio 1990

BERTONI, Carla e Marina Mantani, *"La Legge per la Formazione della Piccola Proprietà Contadina in Provincia di Ravenna" (1948-1975), Annali dell'Istituto Alcide Cervi*, No. 7, 1985.

BINSWANGER, Hans and Miranda Elgin, "What are the Prospects for Land Reform?", in <u>Proceedings of the XX IAAE Conference</u>, Buenos Aires, 1988.

CORSARO, L., *"Il credito agevolato per la formazione della proprietà coltivatrice" in Rivista di Diritto Agrario*, 1974, p. 362 ?

D'ELIA, Costanza, *"Formazione della Proprietà Contadina e Intento Statale in Italia (1919-1975)", La Questione Agraria*, No. 23, 1986.

--   *"Dieci anni di studi sulla proprietà contadina (1974-84)", in Bolletino del Centro Studi per la storia comparata della società rurale in età contemporanea*, 1984.

EUROPEAN PRODUCTIVITY AGENCY of the OEEC, <u>The Small Family Farm, A European Problem</u>, Methods for Creating Economically Viable Units, Paris 1959.

GALLONI, G., *Il rapporto giuridico di bonifica*, Milano, 1964, pp. 173 ff (re: the legal nature and functions of the *Cassa*).

GRAZIANI, C.A., *"Gli aspetti istituzionali della questione fondiaria", in Nuovo Diritto Agrario*, 1982, 3 ss ?

GRILLENZONI, Maurizio, *Il Valore della Terra*, Bologna, Edagricole, 1981.

---   *Mobilita Fondiaria e Proprietà Coltivatrice, Cassa per La Formazione della Proprietà Contadina, Convegno Nazionale,* Faenza, 13 Aprile 1991.

--- , Guido Bazzani and Aldo Bertalozzi, <u>Features of farmland market in Italy and in Emilia-Romagna</u>, Bologna, *Università degli Studi*, (I.E.R.Co.), 1989.

KISH, George, <u>Land Reform in Italy</u>, Final Report to the Office of Naval Research, ORA Project 0-28-93, University of Michigan, Ann Arbor, MI, June 1966.

LEVI, G., *L'eredità immateriale*, Torino, Einaudi, 1985.

MARCIANI, G.E., *L'esperienza di Riforma Agraria in Italia*, Rome, SVIMEZ 1966.

MEDICI, Giuseppe, *I Tipi d'Impresa nell'Agricoltura Italiana*, Rome, INEA 1951.

---<u>Land Property and Land Tenure in Italy</u>, Bologna, Edagricole 1952.

MONTANARI, Arianna, *"Evoluzione Fondiaria e Impresa Coltivatrice"* in <u>*La Riforma Fondiaria: Trent'anni dopo,*</u> Milano, Franco Angeli/INSOR, 1979.

PANATTONI, Andrea, *"Il Mercato Fondiario"*, in <u>*Venti Anni di Agricoltura Italiana*</u> *(Scritti in Onore di Arrigo Serpieri e Mario Tofani, a cura della Società Italiana di Economia Agraria e dell'INEA)*, Bologna, Edagricole, ca. 1970.

PIRRE, G., *"La legislazione 'minore' a favore della proprietà contadina nel periodo 1948-60"*, in <u>*Sociologia urbana e rurale*</u>, 1983, No. 10-11.

PONTOLILLO, Vincenzo, <u>Agricultural Credit in Italy Over the Last Decade</u>, Paper prepared for the "Seminar for Selected Countries of the Near East Region and the Mediterranea Basin", Rome 29 January-9 February 1973, Bank of Italy, Servizio Studi,

RAMELLA, F. *Terra e Telai*, Torino, Einaudi, 1985.

ROSI, Mario, *"Recenti variazioni nella distribuzione della proprietà fondiaria"*, Cap. IX in INEA, <u>*La Distribuzione della Proprietà Fondiaria in Italia*</u>, *a cura di Giuseppe Medici*, Roma 1956.

SCHIRINZI, Gualtiero, *"La Mobilità delle Aziende: Nuovi Contributi di Studio"*, <u>*Rivista di Economia Agraria*</u>, No. 4, Dic. 1979.

SHEARER, Eric B., "Italian Land Reform Reappraised", <u>Land Economics</u>, Vol. XLIV No. 1, February 1968.

VANZETTI, Carlo, *"Il Mercato Fondiario in Italia"*, <u>*Giornale degli Economisti*</u>, 3-4, 1965.

ZIZZO, Nino, *"Componenti Congiunturali e Storiche nella Dinamica del Mercato Fondiario"*, <u>*Annali del Mezzogiorno*</u>, Vol. 1, 1961.

ANNEX: Figures

Figure 3-1. Land purchases for the formation or enlargement of family farms under Law 590, by social classes, 1966-75

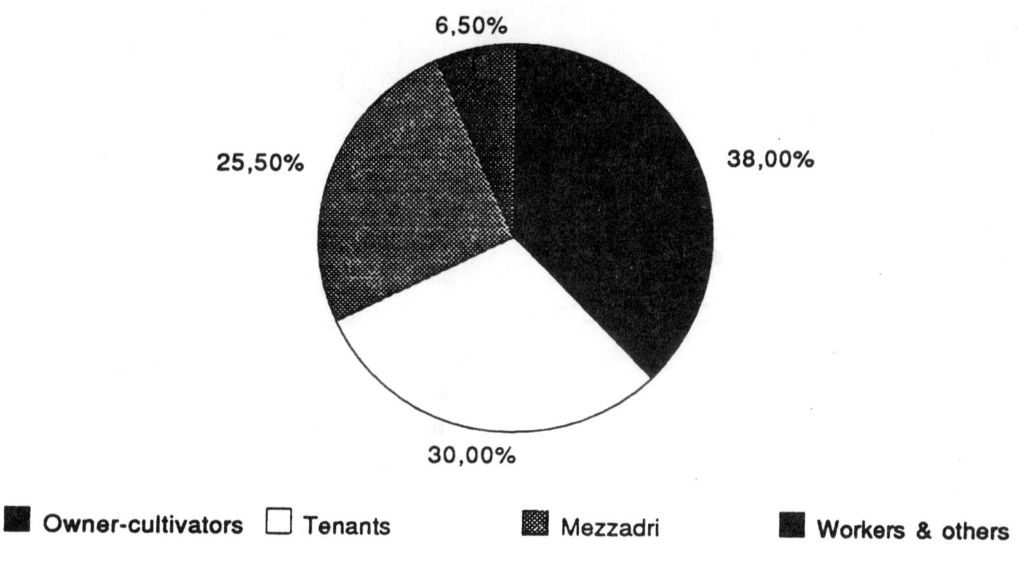

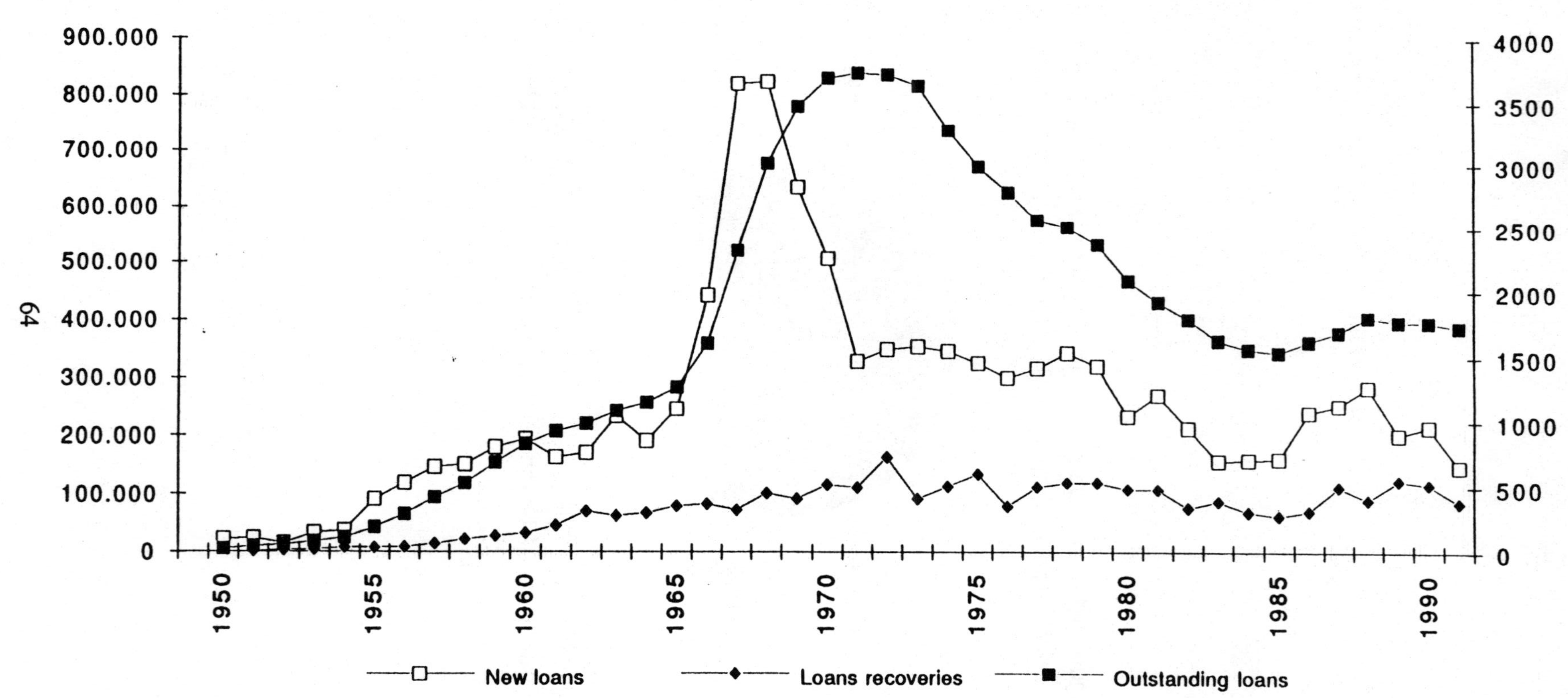

Figure 3-2. Lending and loan recoveries for formation or enlargement of family farms, 1950-91
(left scale: loans and recoveries in million 1991 lire; right scale: outstanding loans in billion 1991 lire)

64

Figure 3-3. Area encompassed by subsidized land market transactions, 1950-91 (hectares)

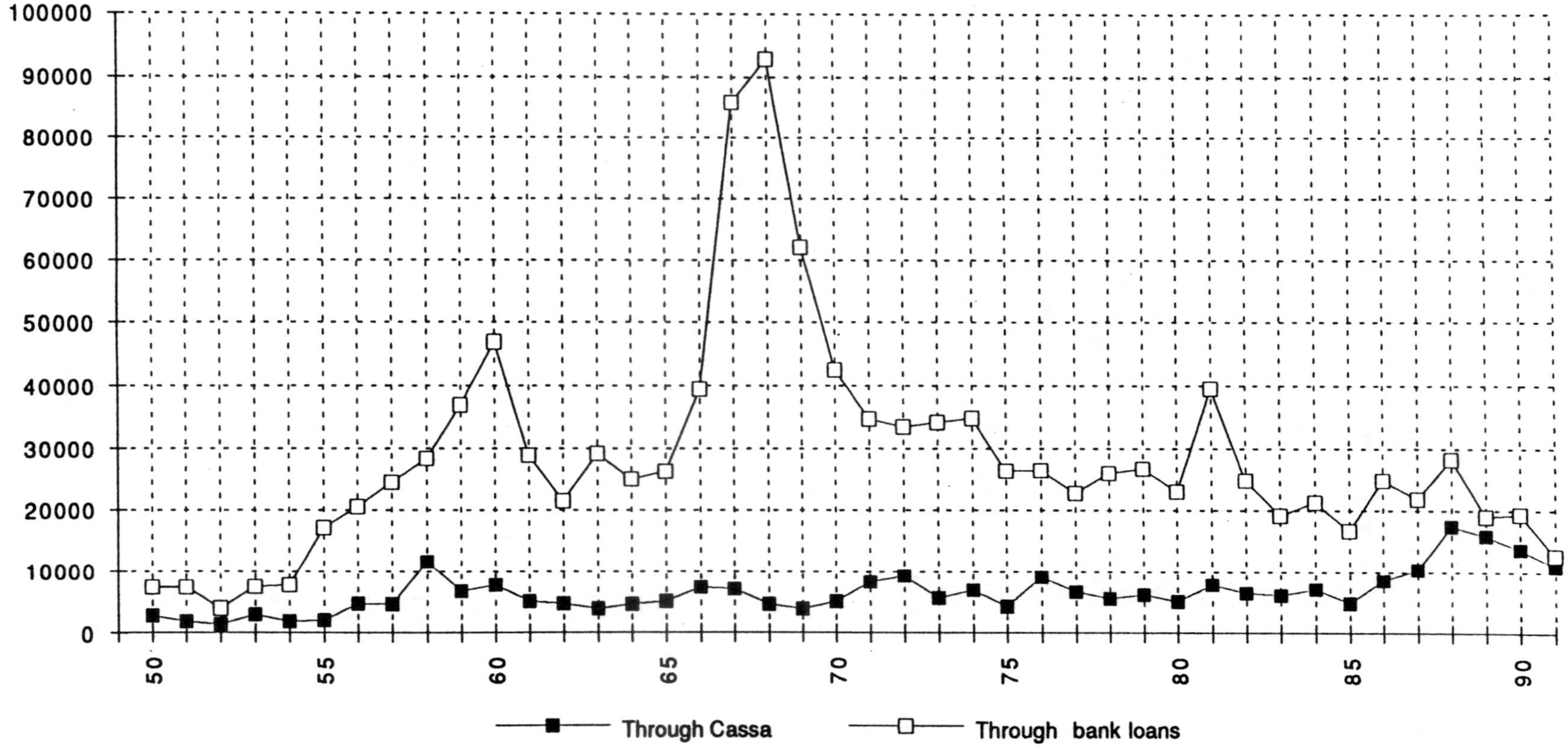

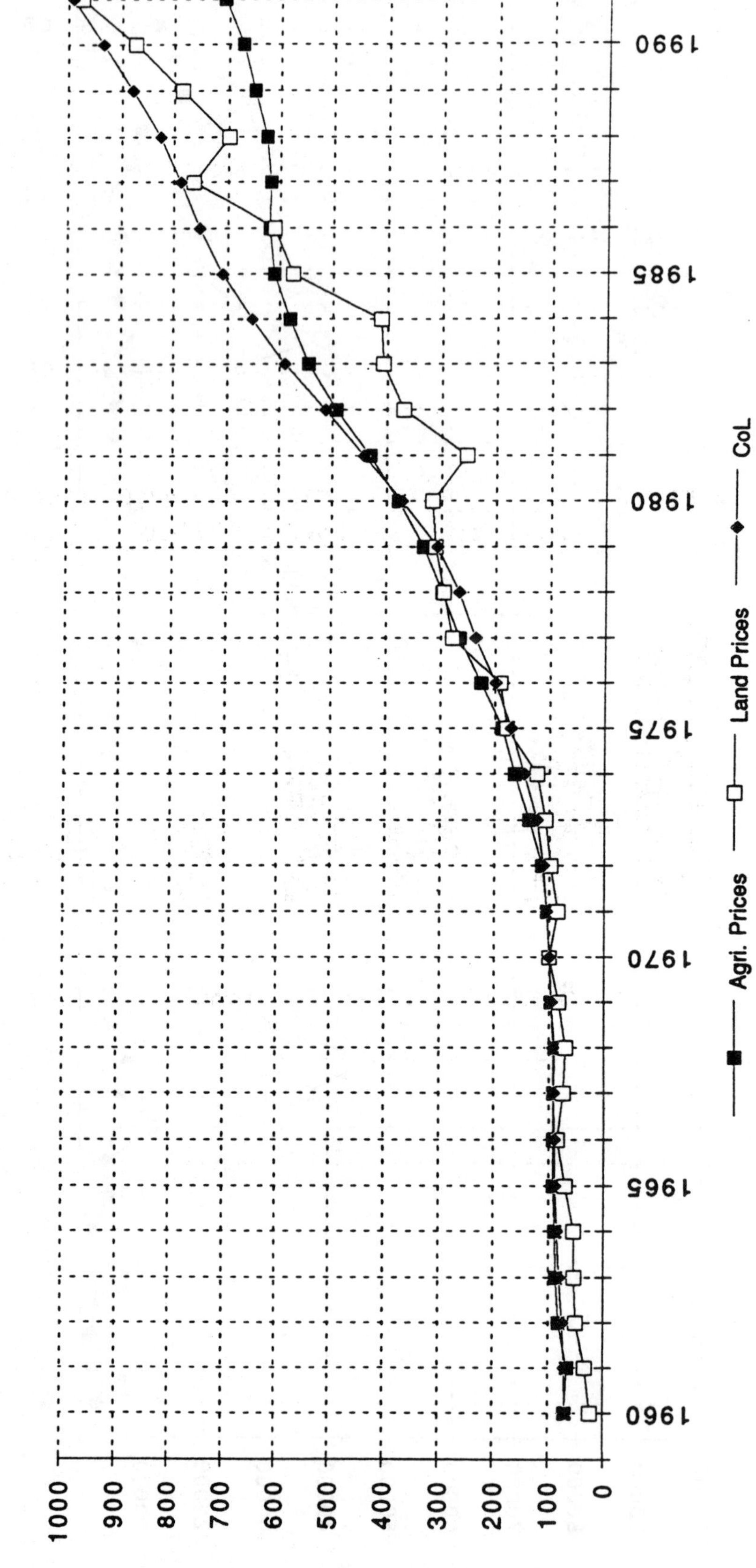

Figure 4-1. Indexes of agricultural prices, farmland prices paid by Cassa and cost of living, 1960-91 (1970 = 100)

Figure 4-2. Ratio of farmland prices to farm wages in three
regions, for selected years, 1957-90

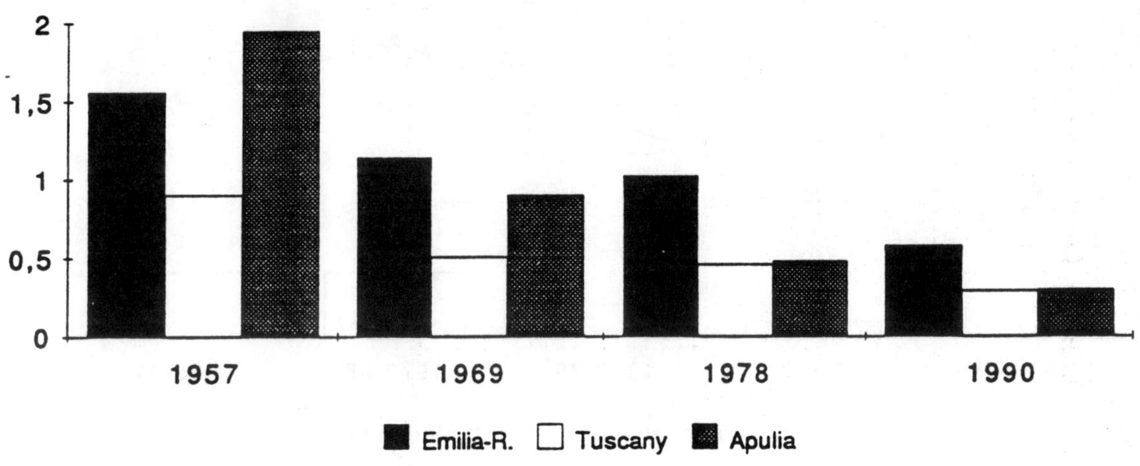

Figure 5-1. Average market rates for long-term loans and Cassa
rates, selected periods, 1948-91

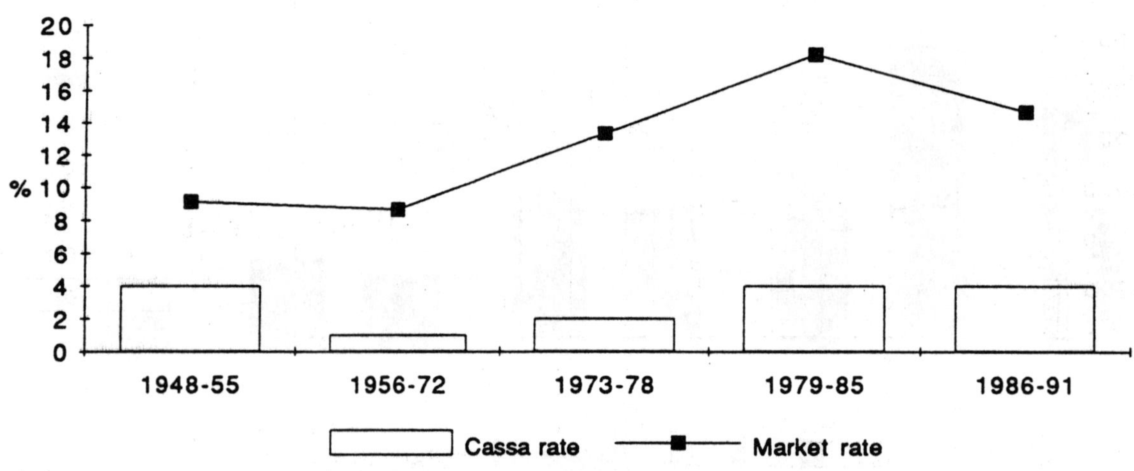

Figure 6-1.   Early repayments by CASSA borrowers, 1972-90
(Families and hectares involved)

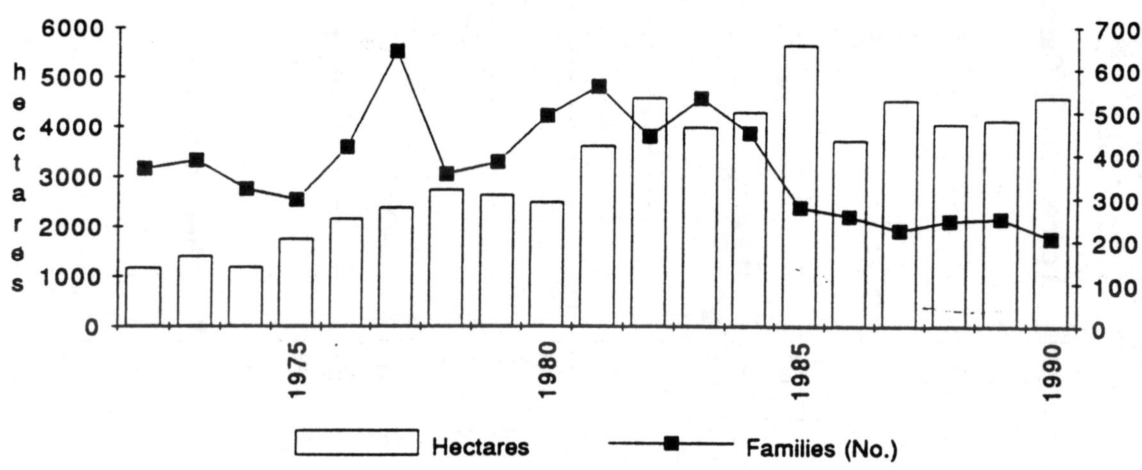

Figure 6-2. Current and real farmland prices paid by Cassa, 1960-61 (real values in lire of 1991)

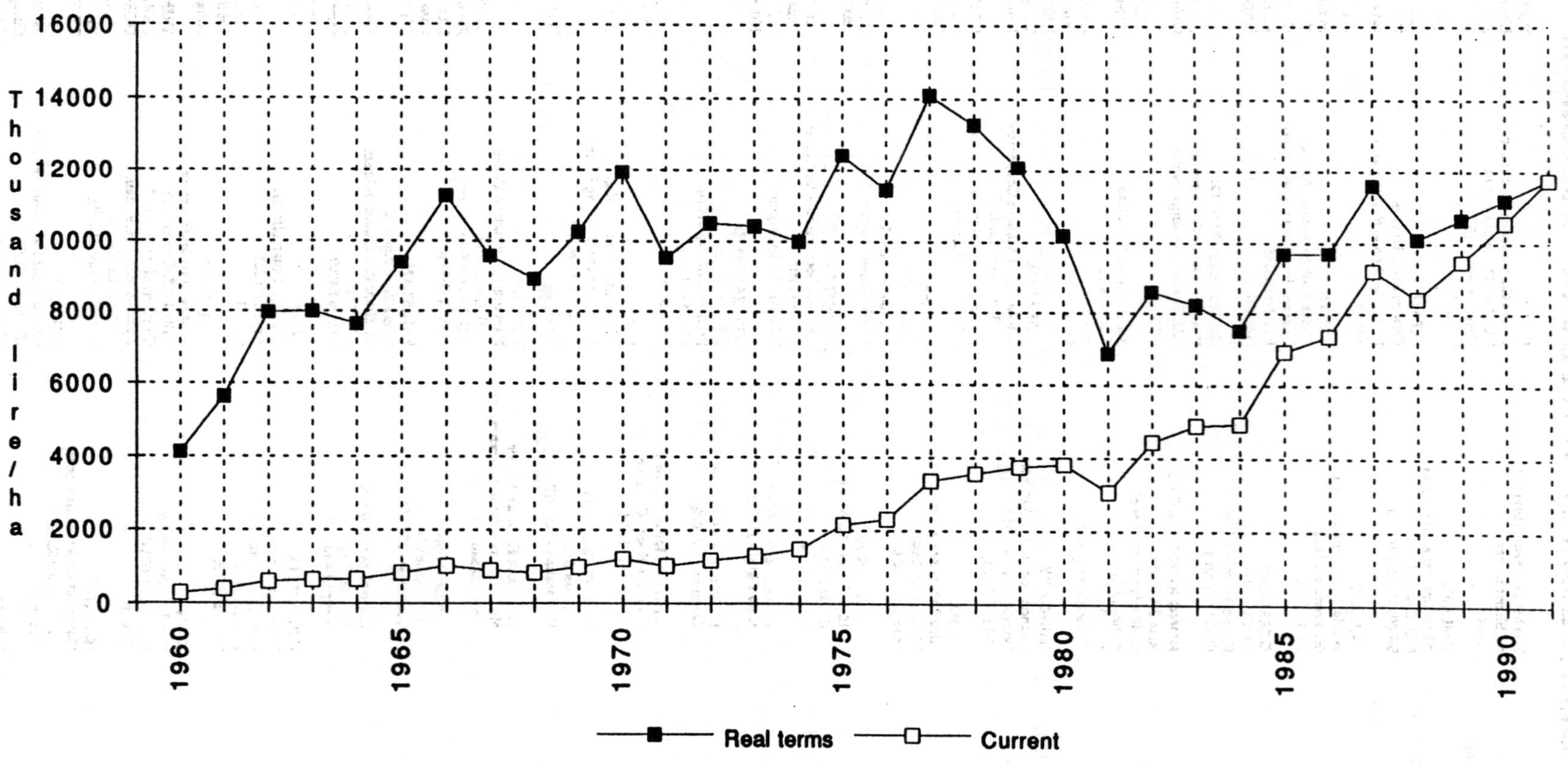

Distributors of World Bank Publications

ARGENTINA
Carlos Hirsch, SRL
Galeria Guemes
Florida 165, 4th Floor-Ofc. 453/465
1333 Buenos Aires

**AUSTRALIA, PAPUA NEW GUINEA,
FIJI, SOLOMON ISLANDS,
VANUATU, AND WESTERN SAMOA**
D.A. Information Services
648 Whitehorse Road
Mitcham 3132
Victoria

AUSTRIA
Gerold and Co.
Graben 31
A-1011 Wien

BANGLADESH
Micro Industries Development
   Assistance Society (MIDAS)
House 5, Road 16
Dhanmondi R/Area
Dhaka 1209

   Branch offices:
   Pine View, 1st Floor
   100 Agrabad Commercial Area
   Chittagong 4100

BELGIUM
Jean De Lannoy
Av. du Roi 202
1060 Brussels

CANADA
Le Diffuseur
151A Boul. de Mortagne
Boucherville, Québec
J4B 5E6

Renouf Publishing Co.
1294 Algoma Road
Ottawa, Ontario
K1B 3W8

CHILE
Invertec IGT S.A.
Av. Santa Maria 6400
Edificio INTEC, Of. 201
Santiago

CHINA
China Financial & Economic
   Publishing House
8, Da Fo Si Dong Jie
Beijing

COLOMBIA
Infoenlace Ltda.
Apartado Aereo 34270
Bogota D.E.

COTE D'IVOIRE
Centre d'Edition et de Diffusion
   Africaines (CEDA)
04 B.P. 541
Abidjan 04 Plateau

CYPRUS
Center of Applied Research
Cyprus College
6, Diogenes Street, Engomi
P.O. Box 2006
Nicosia

DENMARK
SamfundsLitteratur
Rosenoerns Allé 11
DK-1970 Frederiksberg C

DOMINICAN REPUBLIC
Editora Taller, C. por A.
Restauración e Isabel la Católica 309
Apartado de Correos 2190 Z-1
Santo Domingo

EGYPT, ARAB REPUBLIC OF
Al Ahram
Al Galaa Street
Cairo

The Middle East Observer
41, Sherif Street
Cairo

FINLAND
Akateeminen Kirjakauppa
P.O. Box 128
SF-00101 Helsinki 10

FRANCE
World Bank Publications
66, avenue d'Iéna
75116 Paris

GERMANY
UNO-Verlag
Poppelsdorfer Allee 55
D-5300 Bonn 1

HONG KONG, MACAO
Asia 2000 Ltd.
46-48 Wyndham Street
Winning Centre
2nd Floor
Central Hong Kong

HUNGARY
Foundation for Market Economy
Dombovari Ut 17-19
H-1117 Budapest

INDIA
Allied Publishers Private Ltd.
751 Mount Road
Madras - 600 002

   Branch offices:
   15 J.N. Heredia Marg
   Ballard Estate
   Bombay - 400 038

   13/14 Asaf Ali Road
   New Delhi - 110 002

   17 Chittaranjan Avenue
   Calcutta - 700 072

   Jayadeva Hostel Building
   5th Main Road, Gandhinagar
   Bangalore - 560 009

   3-5-1129 Kachiguda
      Cross Road
   Hyderabad - 500 027

   Prarthana Flats, 2nd Floor
   Near Thakore Baug, Navrangpura
   Ahmedabad - 380 009

   Patiala House
   16-A Ashok Marg
   Lucknow - 226 001

   Central Bazaar Road
   60 Bajaj Nagar
   Nagpur 440 010

INDONESIA
Pt. Indira Limited
Jalan Borobudur 20
P.O. Box 181
Jakarta 10320

IRAN
Kowkab Publishers
P.O. Box 19575-511
Tehran

IRELAND
Government Supplies Agency
4-5 Harcourt Road
Dublin 2

ISRAEL
Yozmot Literature Ltd.
P.O. Box 56055
Tel Aviv 61560

ITALY
Licosa Commissionaria Sansoni SPA
Via Duca Di Calabria, 1/1
Casella Postale 552
50125 Firenze

JAPAN
Eastern Book Service
Hongo 3-Chome, Bunkyo-ku 113
Tokyo

KENYA
Africa Book Service (E.A.) Ltd.
Quaran House, Mfangano Street
P.O. Box 45245
Nairobi

KOREA, REPUBLIC OF
Pan Korea Book Corporation
P.O. Box 101, Kwangwhamun
Seoul

Korean Stock Book Centre
P.O. Box 34
Yeoeido
Seoul

MALAYSIA
University of Malaya Cooperative
   Bookshop, Limited
P.O. Box 1127, Jalan Pantai Baru
59700 Kuala Lumpur

MEXICO
INFOTEC
Apartado Postal 22-860
14060 Tlalpan, Mexico D.F.

NETHERLANDS
De Lindeboom/InOr-Publikaties
P.O. Box 202
7480 AE Haaksbergen

NEW ZEALAND
EBSCO NZ Ltd.
Private Mail Bag 99914
New Market
Auckland

NIGERIA
University Press Limited
Three Crowns Building Jericho
Private Mail Bag 5095
Ibadan

NORWAY
Narvesen Information Center
Book Department
P.O. Box 6125 Etterstad
N-0602 Oslo 6

PAKISTAN
Mirza Book Agency
65, Shahrah-e-Quaid-e-Azam
P.O. Box No. 729
Lahore 54000

PERU
Editorial Desarrollo SA
Apartado 3824
Lima 1

PHILIPPINES
International Book Center
Suite 1703, Cityland 10
Condominium Tower 1
Ayala Avenue, H.V. dela
   Costa Extension
Makati, Metro Manila

POLAND
International Publishing Service
Ul. Piekna 31/37
00-677 Warzawa

For subscription orders:
IPS Journals
Ul. Okrezna 3
02-916 Warszawa

PORTUGAL
Livraria Portugal
Rua Do Carmo 70-74
1200 Lisbon

SAUDI ARABIA, QATAR
Jarir Book Store
P.O. Box 3196
Riyadh 11471

**SINGAPORE, TAIWAN,
MYANMAR,BRUNEI**
Gower Asia Pacific Pte Ltd.
Golden Wheel Building
41, Kallang Pudding, #04-03
Singapore 1334

SOUTH AFRICA, BOTSWANA
For single titles:
Oxford University Press
   Southern Africa
P.O. Box 1141
Cape Town 8000

For subscription orders:
International Subscription Service
P.O. Box 41095
Craighall
Johannesburg 2024

SPAIN
Mundi-Prensa Libros, S.A.
Castello 37
28001 Madrid

Librería Internacional AEDOS
Consell de Cent, 391
08009 Barcelona

SRI LANKA AND THE MALDIVES
Lake House Bookshop
P.O. Box 244
100, Sir Chittampalam A.
   Gardiner Mawatha
Colombo 2

SWEDEN
For single titles:
Fritzes Fackboksforetaget
Regeringsgatan 12, Box 16356
S-103 27 Stockholm

For subscription orders:
Wennergren-Williams AB
P. O. Box 1305
S-171 25 Solna

SWITZERLAND
For single titles:
Librairie Payot
Case postale 3212
CH 1002 Lausanne

For subscription orders:
Librairie Payot
Service des Abonnements
Case postale 3312
CH 1002 Lausanne

THAILAND
Central Department Store
306 Silom Road
Bangkok

**TRINIDAD & TOBAGO, ANTIGUA
BARBUDA, BARBADOS,
DOMINICA, GRENADA, GUYANA,
JAMAICA, MONTSERRAT, ST.
KITTS & NEVIS, ST. LUCIA,
ST. VINCENT & GRENADINES**
Systematics Studies Unit
#9 Watts Street
Curepe
Trinidad, West Indies

UNITED KINGDOM
Microinfo Ltd.
P.O. Box 3
Alton, Hampshire GU34 2PG
England